F. M Carryl

Butter records of Jersey cows

F. M Carryl

Butter records of Jersey cows

ISBN/EAN: 9783743377295

Manufactured in Europe, USA, Canada, Australia, Japa

Cover: Foto ©berggeist007 / pixelio.de

Manufactured and distributed by brebook publishing software
(www.brebook.com)

F. M Carryl

Butter records of Jersey cows

BUTTER RECORDS

—OF—

JERSEY COWS

COMPILED BY

F. M. CARRYL

Passaic Bridge, N. J.

MAY, 1885.

ERRATA.

First page, first column, 9th line from bottom, for Naomie's, read Naomi's.

Third page, second column, 1st line, for Agnopimoguk, read Aquopimoquk.

Fourth page, first column, 25th line from top, for Minta, read Uinta. Also second column, 2d line from bottom, for Sutea, read Lutea ; 3d line, for Sara, read Lara ; 4th line, for Medrene, read Medrena.

Sixth page, first column, 8th line from top, for 41.00, read 14.00. Also second column, 19th line from top, for Jesse, read Jessie.

Tenth page, second column, 25th line from top, for Jesse, read Jessie.

Eleventh page, second column, 8th line from bottom, for Miami Rose, read Miami Prize.

Thirteenth page, first column, 10th line from bottom, for Princess, read Princess Rose.

Fourteenth page, first column, 32d line from bottom, for Su Lee, read Su Lu. Also second column, 5th line from bottom, for Jesse, read Jessie.

Fifteenth page, second column, 4th line from bottom, for Rolland's, read Roland's.

Eighteenth page, first column, 19th line from top, for Ninette, read Minette. Also 29th line, for Jesse, read Jessie.

Twentieth page, second column, 22d line from bottom, for Katie Bushford, read Katie Bashford.

Twenty-third page, second column, 2d and 4th lines from bottom, for Thornedale, read Thorndale.

Twenty-sixth page, second column, 24th line from bottom, for Jesse, read Jessie.

Twenty-ninth page, first column, 25th line from bottom, for Arawana Foppy, read Arawana Poppy.

Thirtieth page, first column, 23d line from top, for Fancy June, read Fancy Juno.

Thirty-first page, second column, 21st line from bottom, for Oclera, read Ochra.

Thirty-third page, first column, 8th line from top, for Zitella, read Zittella 2d.

Forty-seventh page, second column, 24th and 25th lines, for Daisy, read Dairy.

Forty-eighth page, first column, bottom line, and second column, 1st and 5th lines, for Trudie, read Trenie.

Fifty-second page, first column, 9th line from bottom, for Azelda 3d, read Azelda 2d.

Fifty-fifth page, first column, 21st line from top, for Albena, read Alhena; also 27th line, for Fautnie, read Fantine; also 13th line from bottom, read Lucky Belle 2d.

Fifty-seventh page, first column, 11th line from bottom, for Pet Gifford, read Pet Gilford.

Sixtieth page, second column, 6th line from top, for Albena, read Alhena.

Sixty-fourth page, second column, 11th line from bottom, for Oxali's, read Oxalis.

BUTTER RECORDS OF JERSEY COWS.

In offering this list of butter records to the public, I have tried to have it correct, and believe it to be so. Any errors found in it, if reported to me, as also any tests not included in it, will in the first case be corrected, and in the last, find due place in a future edition. When a name is preceded by a star, the test is a *rated* one of less than 7 days. I will be thankful for criticisms and corrections, and for full reports of new tests, which will be a more certain way of their being in a future edition than trusting to my seeing them in print. The list of sires includes every one that appears, to the fourth degree, in the pedigree of every cow who has made 14 lbs., or better, in 7 days. The index includes every known 14 lb. cow. Those which do not refer back to some page in the book are imported cows whose pedigree is unknown, so that the index is a complete alphabetical list of every 14 lb. cow with her H. R. number and her record. In reading records, the left of the point is pounds, the right is ounces.

F. M. C.

ABE LINCOLN 268.

Son of Dick Swiveller 74 and Diana 672.
1st sire of.

Sylvia 687	15.08

3d sire of

Roland's Bonnie 2d 18,054	19.02

4th sire of

Jennie Dodo H. 14,448	21.08
Hilda A. 2d 11,120	20.00
Hilda D. 6683	18.05
Nibbette 11,625	14.07

ABRAHAM 228.

Son of Commodore 229 and Buttercup 4th 555.
4th sire of

Coronilla 8367	14.09½

ACHMED 2115.

Son of Vespucius 758 and Caramel 2727.
1st sire of

Nellie Gray of Clermont 10,905	14.01

ACTÆON 914.

Son of L'Empereur 461 and Nilsson 2441.
3d sire of

Naomie's Pride 16745 -	15.02

ADMIRAL 372.

Son of Saturn 94 and Bronx 306.
2d sire of

Lerna 3634	15.12
Iola 4627 -	15.02½
Ideal 11,842	14.12½

3d sire of

Lernella 22,322	14.01½

4th sire of

Ideal Alphea 18,755	14.06
Alphetta 16,531	14.02½

ADMIRAL 1337.

Son of Dick Swiveller 74 and Flirt 326.
4th sire of

Woodland Lass 3444	14.00

ADONIS 39.

Son of Ned 20 and Dot 34.
3d sire of

Canto 7194 -	15.12
Rosabel Hudson 5704	15.12
Lady Ives 3d 6740	14.08
Fandango 12,908	14.03

4th sire of

Alfleda 6744 -	16.04
*Myra Overall 10,317	14.00

AGAWAM 597.

Sire on I. of Jersey dam Daisy of Ipswich 598.
3d sire of

Countess of Scarsdale 18,633 -	14.06

4th sire of

Countess of Scarsdale 18,633	14.06

AJAX 541.

Son of Rob Roy 17 and Beauty 804.
2d sire of

Lily of Maple Grove 5079	16.03

3d sire of

Lida Mullin 9198	16.08
Lizzie D 10,408 -	14.00

ALBERT 44.

Son of Jerry 15 and Frankie 17.

1st sire of

Lady Mel 2d 1795	21.00
Couch's Lily 3237	16.09
Lady Love 2d 2212	16.08
Kitty Colt 2213	15.09½
Fragrance 4059	15.03
Lady Brown 4th 6911	14.12
*Brightness 2211	14.00

2d sire of

Lady Gray of Hill Top 6850	18.12
Belle Grinnelle 4073	18.08
Rosa Miller 4333	17.07
Oktibbeha Duchess 4422	17.04
Jersey Cream 3151	17.00
Lucky Belle 2d 6037	16.14
Dusky 2526	16.10
*Kitty Lake 8250	15.08½
Brightness 3d 14,824	15.05
Olie 4133	15.00
Bloomfield Lady 6912	14.12
Aroma 8518	14.07
Susette 4068	14.04
Rarity 2d 7724	14.02
Maggie May 3255	14.02
Creamer 2467	14.01
Pretty 2526	14.00

3d sire of

*Optima 6715	23.11
Tenella 6712	22.01½
Croton Maid 5305	21.11½
Countess Potoka 7496	18.15
Peggy Leah 3997	18.12
Lady Gray of Hill Top 6850	18.12
May Blossom 5657	18.11
Summerline 8001	18.06
Cordelia Baker 8814	17.09
Hepsy 2d 12,008	17.08
Valhalla 5300	17.08
Arawana Queen 5368	16.09
Belle of Paterson 5664	16.06
Œnone 8614	15.14
Edwina 6713	15.13
Valeric 6044	15.13
Fanny Taylor 6714	15.12
Princess Bellworth 6801	15.10½
*Kitty Lake 8250	15.08½
Signalana 7719	15.04
Usilda 2d 6157	15.02½
Aldarine 5301	15.01½
Favorite's Rajah Rex 16,153	15.00
Mary Clover 9998	14.15
Duchess of Argyle 3758	14.13
Louvie 3d 6159	14.13
Lady Gray of Hill Top 2d 14,641	14.12
Jersey Cream 2d 8519	14.12
Bell Rex 11,700	14.10
Princess Rose 6249	14.08
Deborana 4718	14.08
Maggie May 2d 12,926	14.06
Maggie C. 12,216	14.06
Jennie of the Vale 9553	14.06
Jeannie Platt 6005	14.04
Lottie Rex 18,757	14.04
Prince's Bloom 9729	14.03
Pet Rex 20,166	14.02½
Lady Gray of Hill Top 3d 14,642	14.02
Rarity 2d 7724	14.02
Belle Grinnelle 3d 16,503	14.02

4th sire of

Value 2d 6844	25.02½
Fadette of Verna 3d 11,122	22.08½
Fairy of Verna 2d 10,973	20.03¾
Hilda A. 2d 11,120	20.00
Gardinier's Ripple 11,693	19.12½
Tenella 2d 19,521	18.12
Rosy Kate's Rex 13,192	18.08
Cordelia Baker 8814	17.09
Maggie 3d 3321	17.08
Colt's La Biche 6369	17.02½
Maggie Rex 28,623	17.00½
Katie Bushford 15,982	17.00
Polly Clover 7052	16.15
Grinnell Lass 11,859	16.10
Silvoretta 6852	16.09
Olie's Lady Teazle 12,307	16.05
Alfleda 6744	16.04
Gazella 3d 9355	16.03
Genevieve Sinclair 11,167	16.02
Lady Cecelia 24,821	16.01
*Wabash Girl 14,550	16.00
Rupertina 10,409	15.12½
Elsie Lane 13,302	15.12
Orphean 4636	15.07
Lady Gray of Hill Top 2d 14,641	14.12
Phyllis of Hillcrest 9067	14.12
Cowle's Nonesuch 6199	14.12
Roll of Honor 13,610	14.12
Reception 3d 11,025	14.10
Euphorbia 11,229	14.09½
Marpetra 10,284	14.06
Lady Clarendon 3d 17,578	14.05½
Signetilia 16,333	14.03½
Gem of Sassafras 8434	14.03½
Prince's Bloom 9729	14.03
Lady Gray of Hill Top 3d 14,642	14.02
Hurrah Pansy 12,153	14.01½
Sadie's Choice 7979	14.00

ALBERT 2d 1835.

Son of Albert 44 and Lady Ives 2d 4332.

1st sire of

Rosa Miller 4333	17.07
Bloomfield Lady 6912	14.12

ALBION 490.

Sire on I. of Jersey, dam Bonfanti 388.

1st sire of

Patty of Deerfoot 15,321	16.00

2d sire of

Abbie Z. 3d 14,742	17.90
Deerfoot Girl 15,329	15.08
Polly of Deerfoot 15,328	15.00
Dena of Deerfoot 15,325	14.08
Cressy of Deerfoot 15,324	14.00

3d sire of

Roland's Bonnie 2d 18 054	19.02

ALDINE 1136.

Son of Nelusko 479 and Gazelle of Mobile 1735.

1st sire of

Lucky Belle 2d 6037	16.14
Julia Evelyn 6007	15.15½
Duchess Caroline 3d 6039	15.08
Bettie Dixon 4527	15.00
Starkville Beauty 4897	14.00

2d sire of

Armon 10,862	16.13½
Mountain Lass 12,921	14.09
Gilt Edge C. 12,223	14.03½

Minnie Lee 2d 12,741   14.03
Therese M. 8364   -   14.02
     3d sire of
Marpetra 10,284   14.06

ALLEGANY CHIEF 2818.
Son of Signal 1170 and Corolla 4392.
     1st sire of
Gardinier's Ripple 11,693   19.12½
Euphorbia 11,229   ·   14.09½

ALLEGANY COUNT 2031.
Son of Beacon Comet 13th 1281 and Susan Maria 3048.
     2d sire of
Gardinier's Ripple 11,693   19.12½

ALMONT 2789.
Son of Schinchon 1132 and Tina 3057.
     1st sire of
Kitty Potter 9893   18.05

ALPHEUS 1168.
Son of Mercury 432 and Europa 176.
     1st sire of
Crust 4775   15.07½
     2d sire of
Little Torment 15581   23.02½
Niva 7523   -   15.08
Forsaken 7520   15.01

AMADEUS 1043.
Sire on I. of Jersey, dam Katy Didn't 2734.
     2d sire of
Litty 8017 -   14.00

AMIR KHAN 2573.
Son of Rajah 340 and Clytie 618.
     1st sire of
Ramilly 17,075 -   -   14.00

ANTELOPE 1927.
Son of Golden Ball 1474 and Safrano Rose 3676.
     3d sire of
Baby Buttercup 10,888   -   14.00

APIS 1206.
Son of Collamore's Atlantic 739 and Undine 1864.
     1st sire of
Ida Bashan 4725   18.00

APOLLO 108 J. H. B.
Son of Loyal 70 J. H. B. and Rose 1435 J. H. B.
     1st sire of
*Mabel of St. Mary's 8627 -   18.03
     2d sire of
Viva Le Brocq 13,702   17.07
Prize Rose 19,309   -   15.01
Belle Grinnelle 3d 16,503   -   14.02

AGNOPIMOGUK 808.
Son of Suffolk 607 and Lady of Westbrook 2011.
     4th sire of
Katie Bushford 15,982   17.00

ARAB 245.
Son of Nimrod 2d 246 and Garland 851.
     1st sire of
Lara 4306   17.08
     2d sire of
Jeanette Montgomery 5177   20.00
Silenta 17,685   15.10
Silene 4807   14.00
Jule 3640   14.00

ASGARD 1379.
Son of Little Joker 693 and Celestine 2389.
     1st sire of
Typha 5870   16.11

ASTYANAX 389.
Son of Gen. Scott 46 and Big Duchess 58.
     2d sire of
Attractive Maid 16,925 -   16.13

ATHOL 621.
Son of Prince 55 and Isabel 1575.
     3d sire of
Urbana 5597   16.00
     4th sire of
Corn 10,504   16.02

AUTOCRAT 1065.
Son of Yankee 1003 (27 J. H. B.) and Forget-Me-Not 626 J. H. B.
     3d sire of
Nibbette 11,625   14.07

BABYLON 4723.
Son of Jacob 1377 and Echo 2d 5785.
     1st sire of
Countess Lowndes 26,874   17.08

BADEN BADEN 3973.
Son of Duke of Greyholdt 1035 and Moose 4126.
     1st sire of
Miss Baden Baden 14,760   -   14.14½

BALBOA 1244.
Son of Duke of Greyholdt 1035 and Ibex 2724.
     1st sire of
Verbena of Fernwood 9088   15.00
Sunny Lass 6033   -   14.07

BALSORA 2357.
Son of St. Martin 1482 and Bella 5354.
     2d sire of
Almah of Oakland 11,102   16.14
Belle Thorne 13,369   -   ·   14.11

BALTIMORE BOY 837.

Sire on I. of Jersey, dam Violet of Oakland 2080.

1st sire of

Lady Oaks 2d 5246 - -	15.02

2d sire of

Belle of Millford 7445	14.07

BARKER'S DANDY 3758.

Son of Dandy Dinmont 1058 and Dido Hurd 2581.

1st sire of

Kate Daisy 8204	14.04

BARON 289 J. H. B.

Son of Farmer's Glory 5196 (274 J. H. B.) and Perry Farm Maid 178 J. H. B.

1st sire of

Baron's Rosette 25,988	15.04

BARONET 2240.

Son of Lord Lisgar 1066 and Amelia 484.

1st sire of

*Variella 6337	-	24.01
Chamomilla 7552		16.10
*Pulsatilla 7551		16.03
Bonnie 2d 5742	-	14.11½
Minta 5743		14.10

BARONET 307 J. H. B.'

1st sire of

Queen of Ashantee 2d 16,657	-	14.03½

BARNEY 1491.

1st sire of

*Fawnette of Woodstock 3710	-	17.08
Thornedale Belle 5265		14.08

2d sire of

Jeanette Montgomery 5177	20.00
Thornedale Belle 3d 10,459	15.15
Lydia of Libby 11,698 -	15.03

3d sire of

Almah of Oakland 11,102	16.14
Belle Thorne 13,369 -	14.11

BASHAN 32.

Son of Bill 50 and Violet 23.

3d sire of

Lobelia 2d 6650	14.06

4th sire of

Hillside Gem 16,640	20.00
Elsie Lane 13,302	15.12
Lady Ives 3d 6740	14.08
Lilly Cross 13,796	14.03

BASHAN 146.

Sire on I. of Jersey, dam Beauty 309.

3d sire of

Ida Bashan 4725	18.00
Miss Willie Jones 6918	16.04
Lady Penn 5314	16.00

4th sire of

Pyrrha 6100 -	16.14½
Goldthread 4945 -	16.09
Miss Willie Jones 6918	16.04
Gazella 3d 9355	16.03
Corn 10,504	16.02
Zalma 8778 -	15.05
Faustine 10,354	14.14½
Pet Rex 20,166 - - -	14.02½
Nellie Gray of Clermont 10,905	14.01

BASHAN 2D 363.

Son of Bashan 146 and Oceana 635.

3d sire of

Miss Willie Jones 6918	16.04
Corn 10504	16.02
Zalma 8778 -	15.05
Faustine 10,354	14.14½
Pet Rex 20,166	14.02½

4th sire of

Idalene 11,841	15.08½
Marvel 13734,	15.01

BEACON COMET 675.

Son of Comet 130 and Jersey Belle No. 2 1527.

3d sire of

Lady Josephine 11,560 (8 days)	19.02

4th sire of

Gardinier's Ripple 11,963	19.12½
Forsaken 7520	15.01

BEACON COMET 13TH 1281.

Son of Beacon Comet 675 and Zilla 1692.

3d sire of

Gardinier's Ripple 11,693	19.12½

BEAU 586.

Son of Gen. R. E. Lee 208 and Lina Carroll 1497.

2d sire of

*Dora O. 11,703	17.03

3d sire of

*Miskwa 15,472	-` 19.06

BEAUCLERC 1882.

Son of Scion 1033 and Favorita 3198.

1st sire of

Nelida 2d 8227	15.01½

3d sire of

Countess of Scarsdale 18,633	14.06

BECKWITH'S BULL 29.

Son of Santa Claus 30 and Palestine 26.

3d sire of

Rene Ogden 1568	15.00

4th sire of

Medrene 3939	17.12
Sara 4306 -	17.09
Sutea 4563	16.03
Kalmia 4561 .	15.00

BEECHNUT 109.
Son of Blucher 2d 102 and Fanny 72.
2d sire of

Gold Lace 10,726 -	14.13

3d sire of

Dudu of Linwood 8336	16.15
Gold Mark 10,727	14.14

4th sire of

Dora Doon 12,909 -	15.00

BEESWAX 1931.
Son of Wethersfield 966 and Lilly 2578.
1st sire of

Cordelia Baker 8814	17.09
Mary Clover 9998	14.15

3d sire of

Percie 14,937	16.13

BEE'S WING 59, J. H. B.
1st sire of

Daisy 2d 15,761 -	15.08

BEL CALIPH 1432.
Son of Belisario 640 and Calliope 1326.
1st sire of

Mollie Garfield, 12,172	18.07

BELLINI 1017.
Son of Belisario 640 and Enid 1482.
1st sire of

Bellini's Maid 15,170	15.01½
Bellini La Biche 15,091	14.14½
Susie La Biche 3d 15,171	14.06½

BELISARIO 640.
Son of Pilot Boy 488 and Flora 1422.
2d sire of

Mollie Garfield 12,172	18.07
Renalba 4117 -	17.04½
Bellini's Maid 15,170	15.01½
Bellini's La Biche 15,091	14.14½
Magnibel 7976 -	14.12
Susie La Biche 3d 15,171	14.06½

BEN BUTLER OF BOVINA 2024.
Son of Vermont 893 and Bertha 2d 2264.
1st sire of

Pride of Bovina 8050	16.09
Dorothy of Bovina 9373	15.04

BEN OGDEN 1545.
Son of Ben Rajah 795 and Duchess of Ogden 1567.
3d sire of

Atlanta's Beauty 12,949	21.03

BEN RAJAH 795.
Son of Rajah 340 and Eliza 619.
1st sire of

Myrtle of Ridgewood 7858	14.01

2d sire of

Calendine 9415 -	17.09
Dudu of Linwood 8336	16.15
Cosetta 15,991	14.11
Walkyrie 5708	14.01

3d sire of

Dora Doon 12,909 -	15.00
Favorite's Rajah Rex 16,153 -	15.00

4th sire of

Atlanta's Beauty 12,949	21.03

BEN WESTON 3111.
Son of Hamilton 1074 and Etta 1756.
2d sire of

Lottie Rex 18,787	14.04

BERKSHIRE HILLS 1583.
Son of son of Hebe 872 and Thisbe 607.
1st sire of

Gossip 6165 -	16.07

BERLIN PRINCE 3360.
Son of John Rex 2761 and Lady Mel 429.
1st sire of

Genevieve Sinclair 11,167	16.02

BERTIE 267.
Son of Pilot 3 and Fairy 10.
1st sire of

Thisbe 607	15.12

2d sire of

Thisbe 2d 2201 -	19.01½
Lutea 6563	16.03
Kalmia 4561 -	15.00

3d sire of

Kaoli 18,980 -	17.08
Safrano 4568 -	17.08
Mhoon Lady 6560	17.03
Auria 4567 -	16.13
Gossip 6165 - -	16.07
Cenie Wallace 2d 6557	15.04½
Florry Keep 6556 -	14.14
Mountain Lass 12,921	14.09
Bintana 9837	14.03½
Erith 4564	14.00

4th sire of

Volie 19,465	18.01
Taglioni 9182	14.01

BERTRAND 664.
Son of Lancaster 149 and Bertie 1471.
1st sire of

Epigaea 4631 -	14.07

BERTRAM 1883.
Son of Scion 1033 and Favorita 3198.
1st sire of

Lady Josephine 11,560 (8 days)	19.02

BIJOU 65 J. H. B.
1st sire of

Patterson's Beauty 4760	18.00

2d sire of

Bertha Morgan 4770	19.06
Mollie Brown 7831	16.00

3d sire of

Lydia Darrach 4903	17.14

BILL 50.

2d sire of

Hattie 2d 2901 .	41.00

3d sire of

*Clematis 3174	21.00
Rene Ogden 1568	15.00
Abbie Z. 14,002	14.11
Sal Soda 3721 -	14.07
Belle of Ogden Farm 1570 -	14.00

4th sire of

Summerline 8001	18.06
Mirtha 3437 - -	17.13½
Mirth's Blanche 19,572	17.13½
Medrena 3939 - -	17.12
*Fawnette of Woodstock 3710 -	17.08
Abbie Z. 3d 14,742 -	17.00
*Matindy 6670	16.03
Callie Nan 7959	16.02
Corn 10,504	16.02
Orphean 4636	15.07
Thornedale Belle 5265	14.08
Lobelia 2d 6650 -	14.06
Romp Ogden 3d 5458	14.01

BILL, JR., 182.
Son of Bill 50 and Hattie 428.

1st sire of

Hattie 2d 2901	14.00

2d sire of

Sal Soda 3721	14.07

3d sire of

Summerline 8001	18.06
Orphean 4636	15.07

4th sire of

Countess Potoka 7496 -	18.15
Lady Gray of Hill Top 6850	18.12
Belmeda 6229	18.12
Percie 14,937 - -	16.12
Gem of Sassafras 8434	14.03½

BILLING'S BULL 38.
3d sire of

Attractive Maid 16,925	16.13

4th sire of

Dudu of Linwood 8336	16.15
Attractive Maid 16,925	16.13
Canto 7194 -	15.12
Rene Ogden 1568	15.00
Lady Ives 3d 6740	14.08
Fandango 12,908	14.03

BISMARCK 292.
1st sire of

Tilda 3720	15.00
Zina 1434	14.00

2d sire of

Hazen's Bess 7329	24.11
Maggie Rex 28,263	17.00½
Polynia 10,753	16.07
Deborana 4718	14.08

3d sire of

Hazen's Bess 7329	24.11
Hazen's Nora 4791	20.04
Silveretta 5852 -	16.09
Princess Shiela 7297 -	16.04½
Alhena 15,995	16.03
Œnone 8614 .	15.14
Tohira 8400	15.13
Orphean 4636 -	15.07
Champion Chloe 12,255	15.05½
Dairy C. 12,227	15.00½
Coronilla 8367 -	14.09½
Gilt Edge C. 12,223 -	14.08½
Maggie May 2d 12,926	14.06
Maggie C. 12,216 -	14.06
Minnie Lee 3d 12,941	14.03
Therese M. 8364	14.02
Webster's Pet 4103	14.02
Beauty Bismarck 4967	14.01
Jesse Leavenworth 8248 -	14.00

4th sire of

Value 2d 6844	25.02½½
Tenella 2d 19,521	18.12
Belmeda 6229 - -	18.12
Katie Bushford 15,982	17.00
Bell Rex 11,700 -	14.10
Lady Gray of Hill Top 2d 14,641	14.12
Kate Daisy 8204 - - -	14.04
Lady Gray of Hill Top 3d 14,642	14.02
Hurrah Pansy 12,153	14.01½
* Myra Overall 10,317	14.00
Baby Buttercup 10,888	14.00

BISMARCK 1423.
1st sire of

Trudie 2d 4084	15.00

2d sire of

Pyrrha 6100	16.14½
Beauty Bismarck 4967	14.01

3d sire of

Nimble 22,335	14.10

BISMARCK 2d 351.
Son of Bismarck 292, and Creampot 460.
3d sire of

Katie Bushford 15982	17.00

4th sire of

Hillside Gem 16,640	20.00

BLACK IMPERIAL 255.
Son of Derby 253 and Bronx 306.
1st sire of

Zampa 2194 .	18.00

2d sire of

Valma Hoffman 4500	21.09
Maple Leaf 4768 -	14.12

3d sire of

Oak Leaf 4769	17.10
Lady Penn 5314	16.00
Celia Belle 5865	14.03

4th sire of

Goldthread 4945	16.09
Euphorbia 11,229	14.09½

BLACK KNIGHT 1759.
2d sire of

Enid 2d 10,782 - -	14.07½

BLACK PRINCE OF HANOVER 2873.
Son of Rioter 2d 469 and Leda 799.
1st sire of

*Blossom of Hanover 13,655	17.08
*Lanice 13,656	17.08

BLACK ROGER 326.
4th sire of

Panatilla 4778	18.03
Bessie S. 5002	16.00

BLONDIN 1934.
Sire on I. of Jersey, dam Daisy of Jersey 4576.
2d sire of

Gazella 3d 6027	16.03
Atricia 6029	15.03

3d sire of

Countess Coomassie 19,339	15.08½

BLONDIN 3D 1935.
Son of Blondin 1934 and Belle of Jersey 363.
1st sire of

Gazella 3d 6027	16.03
Atricia 6029	15.03

BLUCHER 48.
2d sire of

*Mica 1983	15.12
Myrtle 2d 211	15.12
Copper 1979	15.07
Lilly Cross 13,796	14.03

3d sire of

*Mica 1983	15.12
Myrtle of Ridgewood 7858	14.01

4th sire of

Hillside Gem 16,640	20.00
Belmeda 6229	18.12
Lida Mullin 9198	16.08
Elsie Lane 13,802	15.12
Canto 7194	15.12
Lady Bidwell 10,303	15.12
Gold Lace 10,726	14.13
Little Sister 11,666	14.12
Lady Ives 3d 6740	14.08
Aroma 8518	14.07
Kate Daisy 8204	14.04
Lilly Cross 13,799	14.03
Lizzie D. 10,408	14.00
Lucy Gaines' Buttercup 5059	14.00
Miami Prize 8100	14.00

BLUCHER 2d 102.
Son of Blucher 48 and Belle Bower 206.
1st sire of

Myrtle 2d 211	15.12

2d sire of

Myrtle of Ridgewood 7858	14.01

3d sire of

Belmeda 6229	18.12
Lida Mullin 9198	16.08
Canto 7194	15.12
Elsie Lane 13,802	15.12

Gold Lace 10,726	14.13
Little Sister 11,666	14.12
Lady Ives 3d 6740	14.08
Aroma 8518	14.07
Kate Daisy 8204	14.04
Lucy Gaines' Buttercup 5058	14.00
Lizzie D. 10,408	14.00
Miami Prize 8100	14.00

4th sire of

Dudu of Linwood 8336	16.15
Alfleda 6744	16.04
Pierrot's Picture 12,481	16.00
Pierrot's Lady Hayes 11,672	15.12
Gold Mark 10,727	14.14
Gold Princess 8809	14.12
*Myra Overall 10,317	14.00

BLUE DICK 166.
2d sire of

Joan d'Arc 2163	16.13½

BLUETOOTH 1821.
Son of St. Helier 45 and Silene 4307.
1st sire of

Bessie Bradford 2d 7271	15.02

BOBBY 208 J. H. B.
Son of Vertumnus 161 J. H. B. and Young Rose 43 J. H. B.
1st sire of

Fear Not 2d 6061	16.02

BON TON 1656
Son of Autocrat 1065 and Bonfanti 388.
2d sire of

Nibbette 11,625	14.07

BOX 1011.
Son of Oakland 33 and Buttercup 2d 1100.
3d sire of

Cowle's Nonesuch 6199	14.12

4th sire of

Jennie of the Vale 9553	14.06½

BRANDYWINE 64.
Son of Chieftain 65 and Juno 120.
4th sire of

Magnibel 7976	14.12

BRAVY 1923.
Son of St. Helier 45 and Helve 4565.
1st sire of

Nannie Fitch 9143	14.04

BRAXTON 1715.
Son of Pertinatti 713 and Brunette Lass 1780.
2d sire of

Daisy Brown 12,213	17.06

BRIGAND 1899.
Son of Khedive 1051 and Lenore 2d 712.
2d sire of

Atlanta's Beauty 12,949	21.03

BRIGHT 308 J. H. B.
1st sire of

Moggy Bright 25,891	16.06

BRISTOL CHIEF 1496.
Son of Wethersfield 966 and Judy 691.
1st sire of

Summerline 8001	18.06

BRITON 919.
Son of Royal Rob 598 and Maud of Ipswich 1847.
3d sire of

Countess of Scarsdale 18,633	14.06

4th sire of

Countess of Scarsdale 18,633	14.06

BROKER 873.
Sire on I. of Jersey, dam Nora 389.
1st sire of

Christmas Nannie 4075	19.07

BRONX BASHAN 145.
Son of Bashan 146 and Bronx 306.
2d sire of

Ida Bashan 4725 -	18.00

3d sire of

Gazella 3d 9355	16.03
Nellie Gray of Clermont 10905 -	14.01

4th sire of

Nellie Gray of Clermont 10905	14.01

BROOKSIDE 1104.
Son of Roderick 128 and May 255.
3d sire of

Yellow Locust 10679	14.10½

BROTHER JACK 4042.
Son of Cinnabar 1739 and Phœdra 2561.
1st sire of

Zitella 2d 11,922 -	17.08½
Malope 2d 11,923	15.10

2d sire of

Reality 16,537	15.03½

BRUTUS WOODFORD 703.
2d sire of

Hazalena's Butterfly 10,123	14.00

BROWN PRINCE 85 J. H. B.
Son of Prince of Wales, I. of Jersey, and Browny 113.
2d sire of

Nelly 6456 -	21.00
Fear Not 6059 -	17.03
Coomassie 11,874 -	16.11
Queen of Ashantee 14,554	15.02

3d sire of

Nancy Lee 7618	26.08½
Fear Not 2d 6061 -	16.02

4th sire of

Princess 2d 8046 -	46.12½
Little Torment 15,581	23.02½
Ona 7840 -	20.13
*Punchinello 11,875 -	17.11½
Young Garenne 13,641	17.08
Daisy Brown 12,213 -	17.06
Lady Velveteen 15,771	17.02
Odelle Sales 15,564 -	16.03
Les Cateaux 2d 15,538	16.01
Rose of Oxford 13,469	15.14½
Lady Kingscote 26,085	15.10
Romping Lass 11,021 -	15.00
Lady Vertumnus 13,217	14.10
Blonde 2d 9268	14.04
La Rouge 12,405	14.02
Daisy Queen 9619	14.00
Gazelle 15,961	14.00
Ada Minka 15,562	14.00
Lady Young 16,668	14.00

BROWNY 158 J. H. B.
Son of Tom 77 J. H. B. and Fairy 964 J. H. B.
1st sire of

Miss Browny 7288	16.13
Beauty 7414 - -	15.00
Rosebud of Belle Vue 7702	14.10
Lizzie C. 7713	14.00

2d sire of

Variella of Linwood 10,954	14.01

BUCKSKIN 151.
Son of Prince John 22 and Custard 321.
2d sire of

Nellie 1507	14.02

3d sire of

Beulah of Baltimore 3270	14.06½

4th sire of

Jesse Lee of Labyrinth 5290	14.07

BUFFER 2055.
Son of Lord Monck 304 and Amelia 484.
1st sire of

Pearl of St. L. 5527	14.02
Moss Rose of St. L. 5114	14.00½

2d sire of

Mary Anne of St. L. 9770	36.12½
Mermaid of St. L. 9771	25.13½
Naiad of St. L. 24,965 -	22.02½
Rioter Pink of Berlin 23,665	19.14
Crocus of St. L. 8351 -	17.12
Judith Coleman 11,391	17.05
*Dido Miss 8759 -	17.01
Moth of St. L. 9775	16.02
Aleph Judea 11,389 - -	15.01¾
Coquette of Glen Rouge 17,559 -	15.01½
Honeysuckle of St. Anne's 18,674	14.14

3d sire of

Rose of St. L. 20,426	21.03½
Rioter's Ruth 14,882	14.12
Rioter's Beauty 14,894	14.00

BULLY BRONX 604.
Son of Kearsarge 257 and Bronx 306.
3d sire of

Hazen's Bess 7329	24.11
Goldthread 4945	16.09

BURDELL 1087.
Son of Living Storm 173 and Belle 1225.
1st sire of

Cascadilla 3103	- 15.12

BURGUNDY 201.
Son of Isaac 42 and Mauve 706.
2d sire of

Oxalis 606	- 15.00

3d sire of

Oxalis 2d 15,631	15.00

4th sire of

Rosa of Glenmore 3179	17.10
Embla 4799	17.08

BURNSIDE 1234.
Son of Hughes 954 and Clematis 3174.
1st sire of

Lobelia 2d 6650	14.06

2d sire of

Princess Shiela 7297	16.04½
Lilly Cross 13,796	14.03

3d sire of

Lady Bidwell 13,303	15.12

BURNSIDE 2d 2838.
Son of Burnside 1234 and Laurel 1973.
2d sire of

Lady Bidwell 10,303	15.12

BUSTLER 137 E. H. B.
Son of Bedesman 49 E. H. B. and Bustle, by Rioter, 746.
3d sire of

La Belle Petite 5472	15.08

BUTTERFLY 156.
Sire on I. of Jersey, dam Newton Belle 355.
3d sire of

Roy al Princess 2370	17.12

3d sire of

Royal Princess 2d 12,346	17.12

BUTTERMAKER 3098.
Son of Ralph Guild 1917 and Sugar 4940.
1st sire of

Aroma 10,862	16.13½

BUTTER PRINT 1863.
Son of Nelusko 479 and Mollie Horton 1734.
1st sire of

Valerie 6044	- 15.13

BUTTER STAMP 101 J. H. B.
Son of Trust 162 J. H. B. and Sultane 7 J. H. B.
1st sire of

Butter Star 7799 -	18.04½
Maid of Five Oaks 7178	15.04

BYRON 279.
Son of Mark Tapley 270 and Betty 683.
2d sire of

Jersey Queen of Barnet 4201 A.	
H. B.	19.12
Roland's Bonnie 2d 18,054	19.02
Snowdrop F. W. 16,948	14.08

CADMUS 4.
2d sire of

Rose of Glenmore 3179	17.10
Embla 4779	17.08

3d sire of

Rosebud of Allerton 6352	19.12
Violet of Glencairn 10,221	14.04
Naomi Cramer 8628 -	14.00

4th sire of

Rosebud of Allerton 6352	19.12

CAEN 2317.
Son of Alpheus 1168 and Countess of Normandy 2675.
1st sire of

Niva 7523	- 15.08

CÆSAREA 214 J. H. B.
2d sire of

Royal Princes 2d 12,346	17.12

CALIPH 1618.
Son of Normandy 1046 and Rosette of Staatsburg 3008.
1st sire of

*Myra Overall 10,317 -	14.00

CALLIS 1696.
Son of Don 611 and Calliope 1326.
1st sire of

Callie Nan 7959	16.02

CAMERLENGO 3012.
Son of Signal 1170 and Maiden of Jersey 2736.
1st sire of

Lady Clarendon 3d 17,578	14.05½

CAMERON 239.
Son of Blucher 2d 102 and Calla 2d 415.
2d sire of

Lucy Gaines' Buttercup 5058	14.00

CANTATOE 1063.
Son of Willie Boy 434 and Daisy 2d 2784
2d sire of

Fair Starlight 1745	17.07½

CARDINAL OF ROSLAND 3335.
Son of Majestic 2d 1201 and Amelia of Greenwood 2139.
1st sire of

*Miskwa 15,472	19.06
*Dora O. 11,703	17.03

CAPT. DARLING 535.
Son of Prince of Jersey 66 and Jersey Queen 1410.

3d sire of

Lady Mel 2d 1795	21.00
Champion Chloe 12,255	15.05½
Bounty 1606 -	14.00
Lady Brown 438	14.00
*Brightness 2211 -	14.00
*Allen's Fawnette 3722	14.00

4th sire of

Value 2d 6844	25.02½½
*Optima 6715	23.11
Peggy Leah 3097	18.12
Kitty Potter 9893	18.05
Maggie 3d 3221	17.08
Dimple 3248 -	16.11
Cascadilla 3103	15.12
*Filbert 3630 -	15.12
*Kitty Lake 8250	15.08½
Romp Ogden 2d 4764	15.05
Brightness 3d 14,824	15.05
Lady Brown 4th 6911	14.12
Silver Belle 4313 - -	14.00
*Churchill's Betsey 4105	14.00

CARELESS BOY 1297.
Son of Sam Weller 271 and Diana 672.

1st sire of

Hennie 3335	15.00

CARTER'S JERRY 3029.
Son of Blucher 2d 102 and Creampot 2d 6738.

2d sire of

Kate Daisy 8204 -	14.04

CARLO 5559.
Son of Hero 126 J. H. B. and Pretty Maid 1493 J. H. B.

1st sire of

Belle Dame 2d 22,043	15.03
Carlo's Fanny 14,951	14.00

CASH BOY 2248.
Son of Rex 1330 and Dido of Middlefield 3416.

1st sire of

Rosy Kate's Rex 13,192	18.08
Maggie Rex 28,623	17.00½

CASTOR 686.
Son of Hudson 116 and Tote 867.

2d sire of

Spring Leaf 5796	14.00

3d sire of

Belle of Vermillion 8798	15.02
Lillian Mostar 10,364	14.03

CATONO 3761.
Sire on I. of Jersey, dam Ona 7840.

1st sire of

Elsie Lane 13,302	15.12

CECCO 1678.
Son of Mercury 432 and Ceres 427.

1st sire of

Ceccola 13,608	16.13
Idaletta 11,843	15.14½
Idalene 11,841	15.08½
Ideal 11,842	14.12½

2d sire of

Ideal Alphea 13,755	14.06

CHAMPION OF AMERICA 1567.
Son of May Boy 705 and Pansy 1019.

1st sire of

Silveretta 5852 -	16.09
Princess Shiela 7297	16.04½
Tobira 8400 - -	15.13
Champion Chloe 12,255	15.05½
Dairy C. 12,227	15.00½
Coronilla 8367 -	14.09½
Maggie May 2d 12,926	14.06
Maggie C. 12,216	14.06
Gilt Edge C. 12,223	14.03½
Minnie Lee 2d 12,941	14.03
Therese M. 8364	14.02
Webster's Pet 4103 -	14.02
Jesse Leavenworth 8248	14.00

2d sire of

Lady Gray of Hill Top 2d 14,641	14.12
Bell Rex 11,700 - -	14.10
Lady Gray of Hill Top 3d 14,642	14.02
Hurrah Pansy 12,153	14.01½
Baby Buttercup 10,888 - -	14.00

3d sire of

Hillside Gem 16,640	20.00
La Pera 2d 13,404	14.08

CHAMPION OF HILLTOP 1839.
Son of Champion of America 1567 and Kate Nickleby 3100.

1st sire of

Lady Gray of Hilltop 2d 14,641	14.12
Lady Gray of Hilltop 3d 14,642	14.02

CHAMPION'S SON 3286.
Son of Champion of America 1567 and William's Birdie 4659.

1st sire of

Baby Buttercup 10,888	14.00

CHALLENGER 376.
Son of Lord Lonsdale 305 and Jessie 980.

3d sire of

Mary Clover 9998	14.15

4th sire of

Mirth's Blanche 19,572	17.18½
Polly Clover 7052	16.15

CHARLESTON 1.
Sire on I. of Jersey, dam Princess 836.

1st sire of

Wybie 595 -	17.04

2d sire of

Chenie 4570	16.00
Thisbe 607	15.10
Ma Belle 4942	15.00
Adina 1942 -	14.04
Belle of Ogden Farm 1570	14.00

3d sire of

Thisbe 2d 2201		19.01½
Mamie Coburn 3798		17.08
Chamomilla 7552		16.10
*Belle of Inda 3867		15.01½
Ma Belle 4942		15.00
Adina 1942	-	14.04
Rose of Hillside 3866	-	14.08½
Maggie May 3255		14.02½
Gilt Edge 2d 4420		14.00

4th sire of

Hazen's Nora 4791	-	20.04
Rosebud of Allerton 6352		19.12
Lucky Belle '2d 6037		16.14
Gossip 6165	-	16.07
Julia Evelyn 6007		15.15½
Valerie 6044 -	-	15.13
Arawana Buttercup 6052		15.05
Arawana Poppy 6053	-	15.02
Bettie Dixon 4527	-	15.00
Florry Keep 6556	-	14.14
Pride of the Hill 4877		14.08
Belle of Milford 7445		14.07
Lillian Morsar 10,364	-	14.03
Flamant 11,270	-	14.02
Elmora Mostar 15,955		14.00

CHARLIE KITTREDGE 1247.

Son of Joseph L. 148 and Countess Kittredge 2592.

1st sire of

Gold Princess 8809	14.12

CHELTEN DUKE 924.

Son of Pilot, Jr., 141 and Duchess 101.

2d sire of

Jaquenetta 10,958	-	- 14.06

CHELTENHAM 80.

Son of Earl 81 and Juno 120.

2d sire of

Silver Rose 4753	.	- 16.14

3d sire of

Sultana 2d 11,798 -	15.04
Woodland Margaret 6215	14.10½
Opaline 7590 -	14.10

4th sire of

Goldthread 4945	16.09

CHIEF BARON 2984.

Son of Chelten Duke 924 and Black Bess 1788.

1st sire of

Jaquenetta 10,958	14.06

CHIEF JUSTICE 252.

Son of Sam Weller 271 and Dairy Maid 992.

2d sire of

Hilda 2d 11,120	20.00
Hilda D. 6683	18.05

CHIEF JUSTICE 2D 1643.

Son of Chief Justice 252 and Hilda 942.

1st sire of

Hilda D. 6683	18.05

CHIEFTAIN 65.

Son of Prince of Jersey 66 and Snowdrop 100.

4th sire of

Urbana 5597 -	16.00
Bathsheba 2556	14.01

CICERO 7657.

Son of Happy 211 J. H. B. and Fleur de l'Air 12,702.

1st sire of

*Pendule 2d 16,709	-	16.08

CINNABAR 1739.

Son of Matchless 906 and Peredot 2388.

2d sire of

Zittella 2d 11,922	17.08½
Malope 2d 11,923	15.10

4th sire of

Reality 16,537	-	15.08½

CLAIMANT 84 J. H. B.

Son of Lemon 170 J. H. B. and Daphne 1066 J. H. B.

1st sire of

Nancy Lee 7618	26.08½

2d sire of

Bohemian Gipsey 17,452	14.11
Lizzie C. 7713 -	14.00

3d sire of

Variella of Linwood 10,954	14.01

CLARENCE 596.

Son of Prince of Orange 184 and Clara 1530.

2d sire of

Pet Anna 1608	-	14.00

3d sire of

Welma 5942 -	17.08
Elinor Wells 12,068 -	14.00

4th sire of

*Queen of Maple Dale Farm 7036	14.14
Marpetra 10,284	14.06
La Rosa 10,078 -	14.00

CLEMENT 115 (61 J. H. B.)

Son of Willie 12 J. H. B. and Clementine 232 J. H. B.

1st sire of

Joan d'Arc 2163	16.13½
Alice of Salem 5053	14.08

2d sire of

Ochra 2d 11,516	16.06½
Lustre 2062	15.08½
Buttery 3502	14.01
Witch Hazel 1360	14.00
Miami Rose 8100	14.00

3d sire of

Eveline of Jersey 6781		18.06
Conover's Beauty 12,650		18.00
Beeswax 9807	-	17.05
Bella Donna 184 J. H. B.		16.10
Lily of Maple Grove 5079		16.08
Matindy 6670		16.03

Dairy Pride 4th 521 J. H. B.	16.00
Dairy Pride 6th 21,681	16.00
Rose of Oxford 13,469	15.14½
Witch Hazel 4th 6131	15.05½
Atricia 6029	15.03
Dora Doon 12,909 -	15.00
Alice of the Meadows 20,748	14.12
Gold Princess 8809 -	14.12
*Woodland Margaret 6215	14.10½
Opaline 7590	14.10
Gilda 2779	14.06
Denise 8281	14.04½
Litza 6338	14.03
Fandango 12,908 -	14.03
Romp Ogden 3d 5458	14.01
Pixie 4115	14.00

4th sire of

Rosa of Belle Vue 6954	18.07½
Bonnie Yost 7943	18.02
Leoni 11,868 -	18.02
Young Garenne 13,641	17.08
Viva Le Brocq 13,702 -	17.07
Mary Jane of Belle Vue 6956	17.07
Maudine of Elmwood 8718	16.15
Dudu of Linwood 8336	16.15
Lida Mullin 9198	16.08
Gold Trinket 9518 ∶	16.02
Lilly of Burr Oaks 11,001	15.13
Etiquette 4300 -	15.08
Grace Felch 8291	15.00
Romping Lass 11,021	15.00
Rosy Dream 9808	14.13
Caroline 12,019	14.08
Jaquenetta 10,958 -	14.06
Milkmaid of Burr Oaks 9035	14.05
Lucetta 6856 -	14.03
Variella of Linwood 10,954	14.01
*Myra Overall 10,317 -	14.00
Gazelle 15,961	14.00
Lizzie D. 10,408	14.00
Elite 4299	14.00
Bellita 4553	14.00

CLIFF 176.
Son of Dick Swiveller 159 and Fanny 365.
1st sire of

Thisbe 2d 2201	19.01½

2d sire of

Estrella 2631 -	14.12
St Perpetua 2d 5557	14.00

3d sire of

Gossip 6165 -	16.07
Lady Louise 4339	15.00
Gilt 4th 4208	14.00
Ramilly 17,075	14.00
Gilt Edge 2d 4420	14.00

4th sire of

Cenie Wallace 2d 6557	15.04½
Bettie Dixon 4527	15.00
Florry Keep 6556	14.14
Gilt Edge C. 12,223	14.03½
Sasco Belle 13,601	14.00

CLIFFS 290.
Son of Sark 123 and Mollie 370.
4th sire of

Milkmaid of Burr Oaks 9035	14.05

CLIFFORD 286.
Son of St. Helier 45 and Heartsease 508.
1st sire of

Reckless 3569	17.08

2d sire of

Willis 2d 4461 -	16.03
Lebanon Daughter 6106	14.04
Lebanon Lass 6108	14.02

CLIFTON 6.
1st sire of

*Clematis 3174	21.00

2d sire of

*Clematis 3174	21.00
Lobelia 2d 6650	14.06

3d sire of

Princess Shiela 7297 -	16.04½
*Mica 1983	15.12
Lobelia 2d 6650	14.06

4th sire of

Princess Shiela 7297	16.04½
Lady Bidwell 10,303	15.12
Lobelia 2d 6650 -	14.06
Lilly Cross 13,796	14.03

CLIFTONDELL 1117.
Son of Grey Friar 567 and Lady Bowen 354.
2d sire of

Princess Mostar 9700	17.03
Lillian Mostar 10,364	14.03
Elmora Mostar 15,955	14.00

CLIFTON MONARCH 3546.
Son of Duke of Bloomfield 1544 and Zingara 1939.
1st sire of

Elmora Mostar 15,955	14.00

CLIMAX 1249.
Son of Major 883 and Flora 3217.
2d sire of

Cascadilla 3103	15.12

CLIVE 319.
Son of Potomac 153 and Dove 332.
3d sire of

Maudine of Elmwood 8718	16.15
Fancy Juno 6086 -	15' 10
Oitz 8649 -	15.01
Buttery 3502	14.01

4th sire of

Lady Cloud 19,358	16.10
Azelda 2d 7022 - -	15.02
Queen of De Soto 12,318.	14.13
*Myra Overall 10,317	14.00

CLIVE DUKE 1901.
Son of Prize Duke 942 and Welcome Beauty 1268.
1st sire of

Oitz 8649	15.01

CLOVERINE 3510.
Son of Grand Duke Alexis 1040 and Kitty Clover 1113.

Rupertina 10409	15.12½

COCKADE 1979.
Son of Alpheus 1168 and Cocoanut 1694.
1st sire of

Forsaken 7520	15.01

COLONEL 76.
4th sire of

Countess of Lakeside 12,135	19.07
Judy 691 - -	19.00
Countess Micawber 1759	16.08
Lucy Gray 2746	15.13
Helen 3556	15.09
Sylvia 687	15.08
Jersey 3260 -	15.06
Topsey Roxbury 7796	15.00
Angela 1682	14.02

COLONEL CROCKETT 1694.
Son of Tom Dasher 420 and Capatolia 2069.
2d sire of

Value 2d 6844 -	25.02½
Hurrah Pansy 12,153	14.01½

COLLAMORE'S ATLANTIC 739.
2d sire of

Ida Bashan 4725 -	18.00
Daisy of Belhurst 3114	16.08
Gazella 3d 9335	16.03

3d sire of

Sasco Belle 13,601	14.00

4th sire of

Countess Lowndes 26,874	17.08

COLT, JR. 825
Son of Rob Roy 17 and Maggie 2054.
1st sire of

Chloe Beach 3931	14.08

2d sire of

Hepsy 2d 12,008 - -	17.08
Arawana Queen 5368	16.09
Princess Belworth 6801	15.10½
Arawana Buttercup 6052	15.05
Usilda 2d 6157 -	15.02½
Favorite's Rajah Rex 16,153	15.00
Louvie 3d 6159	14.13
Belle Rex 11,700	14.10
Princess 6249 -	14.08
Jeannie Platt 6005 -	14.04
Lottie Rex 18,757	14.04
Pet Rex 20,166	14.02½

3d sire of

Rosy Kate's Rex 13,192	18.08
Maggie Rex 28,623	17.00½
Elsie Lane 13,302 - - -	15.12

4th sire of

Genevieve Sinclair 11,167	16.02

COLUMBIAD 534.
2d sire of

Alluring 5541 -	19.05
Aspirant 9272 -	14.07
Rose of Rose Lawn 9365	14.06
Deoine 6343	14.03

3d sire of

Dark Cloud 9364	15.03½

COLUMBIAD 2D 1515.
Son of Columbiad 534 and Celestia 1898.
1st sire of

Alluring 5541	19.05
Aspirant 9272 -	14.07
Rose of Rose Lawn 9365	14.06
Deoine 4363	14.03

COMET 86.
Son of Bryce 12 and Angelina Baker 13.
3d sire of

Pansy of Bellwood 2d 8904	18.00

COMET 130.
Sire on I. of Jersey, dam English Beauty 449.
1st sire of

Abbie Z. 14,002	14.11

2d sire of

Abbie Z. 3d 14,742	17.00
Silversides 3857	14.03

3d sire of

Nellie Maitland 4450	16.00
Etiquette 4300 - -	15.08
Irene of Short Hills 5137	14.06½
Silversides 3857	14.03
Audrey 1447 -	14.00
Vivalia 12,760	14.00
Elite 4299	14.00

4th sire of

Lady Josephine 11,560 (8 days)	19.02
Siloam 17,623 - - -	18.10
Mamie Coburn 3798	17.08
Silvia Baker 7893 -	16.04
Countess Coomassie 19,339	15.08½
Irene of Short Hills 5137 -	14.06½

COMET 223.
Son of Napoleon 225 and Duchess 550.
1st sire of

Judy 691	19.00

2d sire of

Tilda 3720	15.00

3d sire of

Peggy Leah 3097 - -	18.12
White Clover Leaf 4512	17.15
Duchess of Argyle 3758	14.13
Alice of the Meadows 20,748	14.12
Topsey K. 22,769 -	14.00

4th sire of

Miss Blanche 2515 (10 days)	20.09
Polly Clover 7052 -	16.15
Alice of the Meadows 20,748	14.12

COMMODORE 56.
1st sire of

Effie 523	19.11

2d sire of

Grace 2d 919	20.00
Lara 4306	17.08
Oxalis 606	15.00
Heartsease 503	15.00

3d sire of,

Reckless 3569	17.08
Haddie 921 -	16.00
Silenta 17,685	15.10
Oxalis 2d 15,631	15.00
Memento 1913 -	14.05
Eureka McHenry 8341	14.00
Pansy 602	14.00
Silene 4307	14.00

4th sire of

Mirth's Blanche 19,572	17.13½
Mirtha 3437 -	17.13½
Rosa of Glenmore 3179	17.10
Reckless 3569	17.08
Embla 4799 -	17.08
*Dora O. 11,703	17.03
Allie Minka 2982	14.06½
Memento 1913	14.05
Cigarette 2849	14.04
Helve 4565 -	14.00
Naomi Cramer 8628	14.00
Muezzin 3670 -	14.00

COMMODORE 229.
2d sire of

Angela 1682	14.02

3d sire of

Duchess of Bloomfield 3653	20.00½
Su Lee 4705	17.15
Vixen 7591	17.06
Pattie Mc. 3d 4754	16.08
*Pulsatilla 7551	16.03
Letitia 3977	15.05
Bathsheba 2556	14.01

4th sire of

Roonan 5133	18.12
*Variella 6337	18.03¾
Lara 4306 -	17.08
Pattie Mc. 3d 4754	16.08
Urbana 5597 -	16.00
Kate Gordon 8387	15.15
Lorella 12,913	14.07
Lucetta 6856	14.03
Litza 6338 - -	14.03
Variella of Linwood 10,954	14.01
Pixie 4115	14.00

COMMODORE NUTT 36.
Son of Emperor 2d 37 and Mignonette 6.
2d sire of

Olie 4133 -	15.00
Duchess of Argyle 3758 -	14.13
Zina 1434 -	14.00

3d sire of

*Optima 6715	23.11
Polynia 10,753	16.07
Alhena 15,995 - -	16.03
Cascadilla 3103 -	15.12
Romp Ogden 2d 4764	15.05
Deborana 4718	14.08
* Churchill's Betsy 4105	14.00

4th sire of

Hazen's Bess 7329	24.11
Hazen's Nora 4791 -	20.04
Olie's Lady Teazle 12,307	16.05
Œnone 8614 -	15.14
Webster's Pet 4103 - -	14.02
Lady Gray of Hill Top 3d 14,642	14.02
Romp Ogden 3d 5458 -	14.01
Beauty Bismarck 4967	14.01

COMMODORE ROXBURY 1586.
Son of Roxbury 247 and Boquet 852.
2d sire of

Urbana 5597	16.00

COMPEER 2367.
Son of son of Alphea 562 and Bella Donna 1727.
1st sire of

Forget-Me-Not-O 10,564	15.04
Sadie's Choice 7979	14.00

2d sire of

Signetilia 16,333	14.03½

COMPETITOR 237.
4th sire of

Cascadilla 3103 -	15.12
Mary Clover 9998 - -	14.15
Milkmaid of Burr Oaks 9035	14.05

COMPO BOY 2830.
Son of Signal 1170 and Lucilla 2735.
1st sire of

Gazella 3d 9355 -	16.03
Reception 3d 11,025	- 15.08½

COMUS 54.
Sire on I. of Jersey, dam Diana 77.
1st sire of

Plenty 950	14.08

2d sire of

Zampa 2194	18.00
Haddie 921	16.00
Turquoise 1129	14.03
Nellie 1507 -	14.02

3d sire of

Beulah of Baltimore 3270	14.06½
Allie Minka 2982	14.06½

4th sire of

Valma Hoffman 4500	21.09
Lady Josephine 11,560 (8 days) -	19.02
Conover's Beauty 12,650 -	18.00
Cyrone 4th 480	17.01
Ochra 2d 11,816	16.06½
Cornucopia 3414 -	15.12
Naomi's Pride 16,745	15.02
Bessie Ridgely 8293 -	14.11½
Jesse Lee of Labyrinth 5290 -	- 14.07
Celia Belle 5865	14.03
Buttery 3502 -	14.01
Ada Minka 15,562 -	14.00
Miami Prize 8100	14.00

COMUS 2d 97.
Son of Oscar 98 and Caroline 181.
2d sire of

Cornucopia 3414	15.12

4th sire of

Dark Cloud 9364 - -	15.03½
Rose of Rose Lawn 9365 -	14.06

CONCHA 1397.
Son of Yankee 1003 and Phœbe 5th 2330.
3d sire of

Armon 10,862	16.13½

CONCORD 1405.
Son of Harry 72 and Countess 3d 990.
2d sire of

Moss Rose of W. F. 5194 -	18.08½
Del of Willow Farm 22,464	14.08

3d sire of

Pavon 12,485 .	14.08

CONQUEROR 89 J. H. B.
Son of Welcome 166 J. H. B. and Princesse 769 J. H. B.
1st sire of

Lucilla Kent 8892	15.10

2d sire of

Lactine 10,680 -	17.01½

COOMASSIE 1484.
Son of Wethersfield 966 and Belle 1225.
1st sire of

Deborana 4718	14.08

COSSACK 1159.
Son of Clement 115 and Daffodil 335.
1st sire of

Ochra 2d 11,516 -	16.06½

COUNT 1403.
Son of Czar 380 and Countess 114.
2d sire of

Molly 3554 -	16.00

3d sire of

Moss Rose of Willow Farm 5194	18.08½
Helen 3556	15.09

4th sire of

Moss Rose of W. F. 5194	18.08½
Pavon 12,485	14.08
Meines 3559	14.00

COUNT BISMARCK 732.
Sire on I. of Jersey, dam Lucy 1827.
1st sire of

Myth 2837 -	14.06

CŒUR DE LION 318.
1st sire of

Maud Lee 2416 -	23.00
Cornucopia 3414	15.12
Corinne 707	14.07

2d sire of

Vivalia 12760	14.00

4th sire of

Ochra 2d 11,516	16.06½

CŒUR DE LION 140 J. H. B.
1st sire of

Saragossa 22,019	15.02

COVENTRY 790.
Son of Bill 50 and Cowslip 43.
2d sire of

Abbie Z. 14,002 -	14.11

3d sire of

Abbie Z. 3d 14,742	17.00

4th sire of

Lady Bidwell 10,303	15.12

CROWN PRINCE OF WOODBURN 1211.
Son of Hillhurst 1210 and Sincerity 3198.
3d sire of

Miss Baden Baden 14,760	14.14½

CRITIC 540.
Son of Orange Peel 502 and Cannie 1359.
1st sire of

Gilda 2779	14.06

2d sire of

Lily of Burr Oaks 11,001	15.13
Grace Felch 8291 - -	15.00
Milkmaid of Burr Oaks 9035	14.05

3d sire of

Jenny Dodo H. 14,448 -	21.08

4th sire of

Wakena 19,721	16.00

CZAR 251.
Son of Pilot Boy 488 and Wanda 1423.
1st sire of

Adina 1942	14.04

CZAR 273.
Sire on I. of Jersey, dam Jenny 686.
2d sire of

Sylvia 687	15.08

3d sire of

Christmas Nannie 4075 -	19.07
Abbie Z. 3d 14,742 -	17.00
Countess Micawber 1759	16.08
Patty of Deerfoot 15,321	16.00
Deerfoot Girl 15,329 -	15.08
Polly of Deerfoot 15,328	15.00
Dena of Deerfoot 15,325	14.08
Dolly of Lakeside 10,824 -	14.08
Daisy of Chenango, 18,582	14.07
Gilda 2779 -	14.06
Cressy of Deerfoot 15,324	14.00

4th sire of

Jersey Belle of Scituate 7828	25.03
Rolland's Bonnie 2d 18,054	19.02
*Fawnette of Woodstock 3710	17.08
Lily of Burr Oaks 11,001	15.13
Clara of Lakeside 10,827	15.00

Polly of Deerfoot 15,328   15.00
Dolly of Lakeside 10,824   14.08
Thornedale Belle 5265   -   14.08
Minnie of Scituate 17,829   14.04½

CZAR 880.
Sire on I. of Jersey, dam imported Judy.
3d sire of
Molly 3554   -   16.00
4th sire of
Moss Rose of W. F. 5194   18.08½
Helen 3556   -   15.09
Arawana Poppy 6053   -   15.02
Snowdrop F. W. 16,948   14.08

DAINTY BOY 2955.
Son of Pierrot 636 and Dainty 796.
1st sire of
Hattie Douglass 24,960 -   16.05
2d sire of
Bellini's La Biche 15,091   14.14½

DAIRY BOY 909.
Son of Monmouth 210 and Danaë 3d 1287.
1st sire of
Pattie Mc. 3d 4754   16.08

DANA 3620
1st sire of
Lucilla 3d 9786   -   -   14.02

DAN BUCK, JR., 382.
Son of Jack Frost 31 and Clover 20.
3d sire of
Duchess of Argyle 3758   14.13
Chloe Beach 3931   14.08
4th sire of
Hepsy 2d 12,008   -   17.08
Arawana Buttercup 6052   15.05
Princess Rose 6249   -   14.08
Jesse Leavenworth 8248   14.00

DANDY DINMONT 1058.
Son of May Boy 705 and Hurd's Fairy Queen 2582.
2d sire of
Kate Daisy 8204   14.04

DANIEL DERONDA 2291.
Son of Thorough-bass 564 and Gazella 1880.
1st sire of
Sasco Belle 13,601   14.00

DANIEL WEBSTER 403.
Sire on I. of Jersey, dam Flossa 292.
1st sire of
Royal Princess 2370   17.12
2d sire of
Masena 25,732   -   20.07
Royal Princess 2d 12,346   17.12
Silvia Baker 7893   16.04

DASH OF GLASTONBURY 1959.
Son of Robbins 953 and Lady Dash 2523.
1st sire of
*Kitty Lake 8250   15.08½
2d sire of
Cordelia Baker 8814   -   17.09
Phyllis of Hillcrest 9067   14.12
Roll of Honor 13,610   14.12

DATE 2624.
Son of Dash of Glastonbury 1959 and Dilly 2527.
1st sire of
Phyllis of Hillcrest 9067   14.12

DEACON 293.
2d sire of
Peggy of Staatsburg 2342   14.01½
3d sire of
Miss Baden Baden 14,760   14.14½
4th sire of
Dorothy of Bovina 9373   15.04

DEERFOOT BOY 1926.
Son of Albion 490 and Daisy of Deerfoot 3182.
1st sire of
Abbie Z. 3d 14,742   -   17.00
Polly of Deerfoot 15,328   15.00
Dena of Deerfoot 15,325   14.08
Cressy of Deerfoot 15,324   14.00

DEFIANCE 196.
2d sire of
Clematis of St. L. 5478 -   14.03
Lily of St. L. 5120   14.00
3d sire of
Sweet Briar of St. L. 5481   22.12
Jolie of St. L. 5126 -   -   15.13½
Honeysuckle of St. Anne's 18,674   14.14
Uinta 5743   -   14.10
Cupid of Lee Farm 5997   14.06
Pearl of St. L. 5527   -   14.02
Nordheim Creamer 9758   14.00
4th sire of
Honeymoon of St. L. 11,221   -   20.05½
Coquette of Glen Ronge 17,559   15.01½

DERBY 253.
Son of Jupiter 93 and Lop Horn 630.
2d sire of
Zampa 2194   18.00
3d sire of
Valma Hoffman 4500   21.09
Oak Leaf 4769   -   17.10
Miss Willie Jones 6918   -   16.04
Maple Leaf 4768   14.12
Spring Leaf 5796   14.00
4th sire of
Oak Leaf 4769   17.10
Lady Penn 5314 -   16.00
Euphorbia 11229   -   14.09½
Celia Belle 5865   14.03

DEVILSHOOF 866.
Son of Ontario 865 and Darkness 1345.
1st sire of

Oak Leaf 4769	- 17.10

2d sire of

Euphorbia 11,229	14.09½

3d sire of

Gardinier's Ripple 11,693	19.12½
Countess Lowndes 26,874	17.08

DIAMOND 155.
3d sire of

Sylvia 687	15.08
Hennie 3335 - -	15.00
Daisy of Chenango 18,582 -	14.07

4th sire of

Perces Lee 5538 - -	16.10
Snowdrop F. W. 16,948	14.08
Alice of Salem 5053	14.08

DIAMOND EARL 3116.
Son of Longfellow 818 and Favorita of Queens Co. 2825.
1st sire of

Belle of Vermillion 8798	15.02
Lillian Mostar 10,364	- 14.03

DICK 118.
Son of Prince Albert 119 and Beauty 228.
2d sire of

Jo 5th 280	17.08

DICK 181 J. H. B.
1st sire of

Buckeye Lass 10,355	14.04

DICK 223 J. H. B.
1st sire of

Beauty of Jersey 7850	19.02

DICK 1410.
Son of Count 1408 and Countess 114.
1st sire of

Molly 3554	16.00

2d sire of

Moss Rose of W. F. 5194	18.08½
Helen 3566	15.09

3d sire of

Moss Rose of W. F. 5194	18.08½
Pavon 12,485	14.08
Maines 3559	- 14.00

4th sire of

Meines 3d 7741 - -	20.01
Del of Willow Farm 22,464	14.08
Pavon 12,485 -	14.08

DICK SWIVELLER 74.
Son of Major 75 and Flora 113.
2d sire of

Countess of Lakeside 12,135	- 19.07

Countess Micawber 1759	16.08
Sylvia 687	15.08
Jersey 3260	15.06

3d sire of

Countess of Lakeside 12,135	19.07
Ebon Edith 10,653 -	19.01
Arawana Poppy 6053 -	15.02
Gilda 2779	- 14.06

4th sire of

Jersey Belle of Scituate 7828	25.03
Roland's Bonnie 2d 18,054	19.02
Moss Rose of W. F. 5194 -	18.08½
Lady Love 2d 2212 -	16.08
Lily of Burr Oaks 11,001	15.13
Niva 7523 - -	15.08
My Queen 12,614 -	15.08
Dolly of Lakeside 10,824 -	14.08
Del of Willow Farm 22,464	14.08
Minnie of Scituate 17,829	14.04½

DICK SWIVELLER 159
Son of Sark 123 and Mollie 370.
2d sire of

Thisbe 2d 2201	19.01½

3d sire of

Estrella 2831 -	14.12
St. Perpetua 2d 5557	14.00

4th sire of

Gossip 6165 -	16.07
Lady Louise 4339 -	15.00
Pride of the Hill 4877	14.08
Gilt Edge 2d 4420	14.00
Ramilly 17,075 -	14.00
Gilt 4th 4208	- 14.00

DICK SWIVELLER, JR., 276.
Son of Dick Swiveller 74 and Twilight 977.
1st sire of

Countess of Lakeside 12,135	19.07
Jersey 3260 - ·	15.06

2d sire of

Countess of Lakeside 12,135	19.07
Gilda 2779 -	14.06

3d sire of

Jersey Belle of Scituate 7828	25.03
Lily of Burr Oaks 11,001	15.13
Dolly of Lakeside 10,824	14.08
Minnie of Scituate 17,829	14.04½

4th sire of

Jersey Belle of Scituate 7828	25.03
Jenny Dodo H. 14,448 - ·	21.08
Jersey Queen of Barnet 4201 A.H.B.	19.12
Roland's Bonnie 2d 18,054 - -	19.02
Belle of Scituate 7977	16.00
Lass of Scituate 9555	15.14
Clara of Lakeside 10,827	15.00
Dolly of Lakeside 10,824 -	14.08
Snowdrop F. W. 16,948	14.08
Minnie of Scituate 17,829 - -	14.04½

DIOGENES 177.
Son of McClellan 323 and Diana 672.
2d sire of

Perces Lee 5538 - -	16.10
Snowdrop F. W. 16,948	14.08

3d sire of

Niva 7523 - - -	15.08

DOCTOR H. 2132.
Son of St. Malo, Jr., 733 and Julia 2d 4902·
1st sire of
Lydia Darrach 4903	17.14

DOCTOR SYNTAX 240 E. H. B,
Son of Tindal 874 E. H. B. and Fawn.
3d sire of
Mary Anne of St. L. 9770	36.12½
Ida of St. L. 24,990 -	30.02½
Mermaid of St. L. 9771	25.13½
*Allie of St. L. 24,991	24.00
Naiad of St. L. 12,965	22.02½
Niobe of St. L. 12,969 - -	21.09½
Honeymoon of St. L. 11.221	20.05½
Rioter Pink of Berlin 23,665	19.14
Crocus of St. L. 8351	17.12
Cowslip of St. L. 8349 -	17.12
Brenda of Elmhurst 10,762	17.04½
Ninette of St. L. 9774	17.04
*Dido Miss 8759 -	17.01
Diana of St. L. 6636	16.08
Maggie of St. L. 9776	16.03
Moth of St. L. 9775	16.02
Minnie of Oxford 12,806	16.00
La Belle Petite 5472 -	16.00
Mavourneen of St. L. 9777	15.07
May Day Stoke Pogis 28353	15.03
Nora of St. L. 12962 -	14.07
Jesse Brown of Maxwell 7266	14.07
Cupid of Lee Farm 5997	14.06
Nancy of St. L. 12964	14.05
4th sire of
Rose of St. L. 20,426 ·	21.03½
Carrie Pogis 22,568	15.09
Rioter's Nora 21,778	15.09
Maggie Sheldon 23,583 -	15.03
Rioter's Ruth 14,882	14.12
Rioter's Beauty 14,894	14.00

DODE 3057.
Son of Emilus 2039 and Lizzie Ringling 5813.
1st sire of
Queen of De Soto 12,318	14.18

DOLPHIN 242 E. H. B.
Son of Wapiti 927 E. H. B. and Doll (Dauncey's).
3d sire of
*Blossom of Hanover 13,655	17.08
*Lanice 13,656 -	17.08
Miss Willie Jones 6918 -	16.04
Pride of Corisande 5323	16.00
Gray Therese 5322	16.00
Myra 2d 6289	16.00
Zalma 8778	15.05
Faustine 10,354	14.14½
Silversides 3857 -	14.03
Pet Rex 20,166 -	14.02½
Robinette 7114 - - ·	14.01
Silver Belle 4313	14.00
4th sire of
Colie 8309	18.04
Typha 5870 -	16.11
Fillpail 16,530 -	15.11
Idalene 11,841 -	15.08½
Forget-Me-Not-O. 10,564	15.04

Marvel 13,734 -
Marvel 13,734 -	15.01
Sadie's Choice 7979	14.00

DOLPHIN 2d 468.
Son of Dolphin 242 E. H. B. and Vanity (Dauncey's).
2d sire of
*Blossom of Hanover 13,655	17.08
*Lanice 13,656 -	17.08
Miss Willie Jones 6918	16.04
Pride of Corisande 5323	16.00
Gray Therese 5322 -	16.00
Myra 2d 6289 -	16.00
Zalma 8778 -	15.05
Faustine 10,354 -	14.14½
Silversides 3857	14.03
Pet Rex 20,166	14.02½
Robinette 7114	14.01
Silver Belle 4313	14.00
3d sire of
Colie 8309	18.04
Typha 5870	16.11
Fillpail 16,530	15.11
Idalene 11,841 - -	15.08½
Forget-Me-Not-O. 10,564 -	15.04
Marvel 13,734 -	15.01
Sadie's Choice 7979	14.00
4th sire of
Smoky 13,738	14.09
Signetilia 16,383 - -	14.03½

DOMINO OF DARLINGTON 2459.
Son of Sarpedon 930 and Beauty of Darlington 5736.
1st sire of
Robinette 7114	14.01
2d sire of
*Pedro Alphea 13,889	15.05

DOM PEDRO 2092.
Son of Iron Bank 1120 and Lebanon 2616.
1st sire of
Dom Pedro's Julian 8631	16.00

DON 611.
Son of Duke 610 and Fawn 476.
1st sire of
Rene Ogden 1568 -	15.00
Belle of Ogden Farm 1570 -	14.00
2d sire of
Medrena 3939	17.12
*Matindy 6670	16.03
Callie Nan 7959 -	16.02
Romp Ogden 3d 5458	14.01
3d sire of
Renalba 4117	17.04½
Calypris 5943 -	15.09½
Romping Lass 11,021 -	15.00
Pride of the Hill 4877	14.08
4th sire of
Atlanta's Beauty 12,949	21.03
Medrie Le Brocq 8888 -	14.07
Vivalia 12,760 -	14.00

DON 218.
Son of Earl 219 and Doll 542.
2d sire of
Alice of the Meadows 20,748   14.12

DON PEDRO 127.
Son of McClellan 25 and Katy Darling 242.
3d sire of
Forget-Me-Not-O 10,564 -   15.04
Sadie's Choice 7979 - - - 14.00
4th sire of
Signetilia 16,333   14.08½

DON PEDRO OF BINGHAMTON 2974.
Son of Vernon 1071 and Zodiac 1914.
1st sire of
Attractive Maid 16,925   16.13

DON PIERROT 2876
Son of Pierrot 636 and Dainty 796.
1st sire of
Mineola of Elmarch 8229   15.15

DOTAJAH 2741.
Son of Ben Rajah 795 and Dot 34.
2d sire of
Favorite's Rajah Rex 16153   15.00

DOUBLE PRIZE 3117.
Son of Beauclerc 1882 and Jewel Beauty 1668.
2d sire of
Countess of Scarsdale 18,633   14.06

DRUID 299.
Son of Earl 81 and Jewel 336.
3d sire of
Jessie Lee of Labyrinth 5290 -   14.07

DUKE 404.
Son of Jerry 15 and Gipsy 2d 737.
4th sire of
*Pedro Alphea 18,689   15.05

DUKE 610.
Son of Garibaldi 609 and Alice 474.
2d sire of
Rene Ogden 1568 -   15.00
Belle of Ogden Farm 1570 -   14.00
3d sire of
Medrena 3939   17.12
*Matindy 6670   16.03
Callie Nan 7959 -   16.02
Romp Ogden 3d 5458   14.01
Litty 8017 -   14.00
4th sire of
Renalba 4117 -   17.04½
Calypris 5943 -   15.09½
Romping Lass 11,021   15.00
Pride of the Hill 4877   14.08

DUKE 1149.
Son of McClellan 323 and Daisy 2954.
2d sire of
Ebon Edith 10,653 -   19.01

DUKE 1400.
Son of Major 75 and Susy 3551.
3d sire of
Helen 3556 -   15.09
4th sire of
Meines 3559 -   14.00

DUKE 4TH 10 J. H. B.
2d sire of
Eveline of Jersey 6781   18.06

DUKE F. 6134.
Son of Byron 279 and Dazzle 379.
1st sire of
Jersey Queen of Barnet 4201 A. H. B. 19.12
Snowdrop F. W. 16,948 -   14.08

DUKE GLENDALE 1819.
Son of Com. Roxbury 1586 and Fawn 850.
1st sire of
Urbana 5597 -   16.00

DUKE PHILIP 843.
Son of St Helier 45 and Duchesse 374.
1st sire of
Avis E. 9714 -   15.14

DUKE OF BLOOMFIELD 1544.
Son of Rioter 670 and Alice Bloomfield 1680.
1st sire of
Princess Mostar 9700   17.03
Princess Bowen 9699   14.12
2d sire of
Elmora Mostar 15,955 -   14.00

DUKE OF BURLINGTON 1639.
Son of Modeste's Maxse 1098 and Favorite of the Elms 1656.
3d sire of
Favorite's Rajah Rex 16,153   15.00

DUKE OF DAFFODIL 1662.
Son of Mack 722 and Daffodil 307.
1st sire of
Alice of the Meadows 20,748   14.12

DUKE OF DARLINGTON 2460.
Son of Sarpedon 930 and Eurotas 2454.
1st sire of
Bomba 10,330 - -   21.11½
Leah Darlington 13,836   15.05½
Nazli 10,327 - -   15.03½
*Lady Golddust 2d 19,861 -   14.01

DUKE OF FRAMINGHAM 15.21.
Sire on I. of Jersey, dam White Rose 3771.
2d sire of

Bet Arlington 8970	18.11
Countess Queen 13,519	18.03
Pet Lee 7993	14.12

DUKE OF GRAYHOLDT 1035.
1st sire of

Duenna's Duchess 5508	16.10
Morlacchi 2725	14.00

2d sire of

Phlox 16,399 - -	21.11
Verbena of Fernwood 9088 -	15.00
Miss Baden Baden 14,760	14.14½
Sunny Lass 6033 -	14.07

DUKE OF LEBANON 1880.
Son of Nye 667 *and Nancy Dawson* 1279.
1st sire of

Blossie Reynolds 6082	16.03½
Home Matron 6707	14.00

2d sire of

Royal Sister 12,457 -	14.11

DUKE OF MAPLEHURST 2290.
Son of Duke of Patterson 1600 *and Clari* 4266.
1st sire of

Litty 8017 -	14.00

2d sire of

*Lenoreisa 16,233	21.00

DUKE OF OAKLAND 1984.
Son of Black Knight 1759 *and Sweetheart* 4196.
1st sire of

Enid 2d 10,782 -	14.07½

DUKE OF OGDEN 210 J. H. B.
1st sire of

Talebearer 24,535 - -	14.08

DUKE OF PATTERSON 1600.
Son of Grand Duke Alexis 1040 *and Kitty Clover* 1113.
2d sire of

Litty 8017	14.00 ·

3d sire of

*Lenoreisa 16,233	21.00

DUKE OF PORTAGE 1270.
Son of Major of Staatsburg 679 *and Lady Palestine* 2769.
1st sire of

Celia Belle 5865 -	14.03

DUKE OF RINDGE 1335.
Son of Magnet 968 *and Maid of the Mist* 2546.
3d sire of

*Wabash Girl 14,550	16.00

DUKE OF SASSAFRAS 2431.
Son of Rossman 1128 *and Kizzie* 3244.
1st sire of

Gem of Sassafras 8434	14.03½

DUKE OF SCITUATE 3623.
Son of Pharos, Jr., 3621 *and Jersey Belle of Scituate* 7828.
1st sire of

Minnie of Scituate 17,829	14.04½

DUKE OF SHREWSBURY 624.
Son of Comet 130 *and Jersey Belle No.* 2 1527.
2d sire of

Irene of Short Hills 5137	14.06½

DUKE OF WELLINGTON 35.
1st sire of

Lassie 1134 -	15.01½

2d sire of

Lucy Gray 2746 -	15.13

3d sire of

Jenny Dodo H. 14,448	21.08

DUKE OF WELLINGTON 608.
Son of Sark 123 *and Jersey Belle* 1526.
1st sire of

Bounty 1606	14.00

2d sire of

*Bonmari 7019	14.00

3d sire of

*Belle of Inda 3867 -	15.01½
Pet Anna 1608 -	14.00

4th sire of

Welma 5942 - -	17.08
Katie Bushford 15,982	17.00
Pet Anna 1608 -	14.00
Elinor Wells 12,068 -	14.00

DUKE OF WEYMOUTH 2515
Son of Lopez 313 *and Lilly Morgan* 4752
3d sire of

Niobe of St. L. 12,969	21.09½

EARL 81.
Son of Monarch 82 *and Europa* 121
1st sire of

Blanche 594	16.00

2d sire of

Ianthe 4562	19.11
* Rosa 663	18.06
Cyrene 4th 480 -	17.01
Belle of Middlefield 1516	16.03
Arietta 5115 -	15.00
Alice of Salem 5053	14.08

3d sire of

Chroma 4572 - -	20.06
Pyrola 4566 - - -	18.06
Cerita of Meadowbrook 5056	17.08

Silver Rose 4753 -	16.14
Pattie Mc. 3d 4754 - -	16.08
Zithey 9184	16.07
Nipheta 9180 -	16.00
Princess Bellworth 6801	15.10½

4th sire of

Beeswax 9807	17.05
Busy Bee 6336 -	16.04
Blossie Reynolds 6082	16.03½
Sultana 2d 11,798	15.04
Charmer 4771	14.12
Renini 9181	14.10½
Opaline 7590	14.10
Trenie 17,770 -	14.10
Jessie Lee of Labyrinth 5290	14.07
Belle of Milford 7445 -	14.07
Queen of Chenango 17,771 -	14.06
Flamant 11,270 -	14.02
Home Matron 6707	14.00

EARL 219.

Son of Santa Ana 221 and Jessie 544

3d sire of

Alice of the Meadows 20,748	14.12

EARL OF BROOKSIDE 1677.

Son of Kago 1353 and Kosi 3431.

2d sire of

La Pera 2d 13,404	14.08

EARL OF WILLOW GLEN 2043.

Son of Majestic 2d 1201 and Queechy 3827.

1st sire of

Mary M. Allison 6308	20.14

ECHO 343.

4th sire of

Miss Willie Jones 6918	16.04

ECLIPSE 1449.

Son of Sweepstakes 682 and Amelia 2d 1730

1st sire of

Nordheim Creamer 9758	14.00

EDDINGTON 460.

4th sire of

Mamie Coburn 3798	17.08
*Belle of Inda 3867 -	15.01½
Rose of Hillside 3866	14.03½
Maggie May 3255	14.02½
Gilt Edge 2d 4420	14.00

EMILIUS 2039.

Son of Jachin 1220 and Emma 2d 2256.

2d sire of

Queen of De Soto 12,318	14.13

EMPEROR 5.

3d sire of

*Sutliff's Rosy 4104 - -	14.00

4th sire of

Hepsy 2d 12,008 - - -	17.08

Olie 4133 - - 15.00
Duchess of Argyle 3758 - 14.13
Zina 1434 - - - - 14.00
*Churchill's Betsy 4105 - 14.00

EMPEROR 287.

Son of Jerry 15 and Eve 2d 734.

1st sire of

Actress 2311 - -	14.00

2d sire of

Pride of Bovina 8050	16.09
Daisy of Clermont 3492 -	14.00

3d sire of

Pride of Bovina 8050	16.09
Dorothy of Bovina 9373	15.04

EMPEROR 338.

Son of Roxbury 247 and Anna 592.

3d sire of

Epigæa 4631 -	14.07

EMPEROR 2D 37.

Son of Emperor 5 and Fawn 25.

2d sire of

*Sutliff's Rosy 4104 -	14.00

3d sire of

Hepsy 2d 12,008 -	17.08
Olie 4133 - - - -	15.00
Duchess of Argyle 3758 - -	14.13
Zina 1434 - -	14.00
*Churchill's Betsy 4105 -	14.00

4th sire of

*Optima 6715 - -	23.11
Silveretta 6852	16.09
Gossip 6165 - - -	16.07
Polynia 10,753 - -	16.07
Princess Shiela 7297	16.04½
Tobira 8400	15.13
Cascadilla 3103 - -	15.12
Champion Chloe 12,255	15.05½
Romp Ogden 2d 4764	15.05
Dairy C. 12,227 - - -	15.01
Coronilla 8867	14.09½
Deborana 4718 -	14.08
Maggie May 2d 12,926	14.06
Maggie C. 12,216 -	14.06
Gilt Edge C., 12,223 -	14.03½
Minnie Lee 2d 12,941	14.03
Webster's Pet 4103	14.02
Therese M. 8864 -	14.02
Jessie Leavenworth 8248	14.00
Gilt 4th 4208 - -	14.00
Churchill's Betsy 4105	14.00

ESSEX OF STAATSBURG 892.

Son of Governor 890 and Eve 5th 312.

2d sire of

Miss Baden Baden 14,760	14.14½

EUCLID 520

Son of Lawrence 61 and Golddrop 222.

1st sire of

Ma Belle 4942	15.00

3d sire of

Attractive Maid 16,925 -	16.13

EXCELSIOR 647.
Son of Ned 523 and Cushing's No. 3 1638.
2d sire of.

Gold Lace 10,726	14.13

3d sire of

Gold Mark 10,727	- 14.14

EXCELSIOR OF JERSEY 949.
1st sire of

Lady Oxford 4860 (10 days)	22.02

2d sire of

Grace Felch 8291 -	15.00
Grace Davy 8292	14.02

3d sire of

*Queen of Maple Dale Farm 7036	14.14
Milkmaid of Burr Oaks 9035	*1405

EXPOUNDER 1148.
Son of Gen. Behm 552 and Lorraine 1435.
2d sire of

Alice Herrick 8778 - -	14.14

EXPRESS 328.
3d sire of

Panatella 4778	18.03
Bessie S. 5002 -	16.00

4th sire of

Belle of Vermillion 8798 -	15.02
Lillian Mostar 10,364	14.08

FAIRFAX 530.
Son of Monmouth 210 and Copia 46.
2d sire of

Beeswax 9807 -	17.05
Busy Bee 6336 -	16.04

3d sire of

Maudine of Elmwood 8718 -	16.15

4th sire of

*Wabash Girl 14,550	16.00

FANCHON'S KING 2637.
Son of Jersey King 879 and Fanchon 2d 1958.
1st sire of

Countess Potoka 7496	18.15

FANCY LEO 2728.
Son of Prize Duke 942 and Fancy Fair 2858.
2d sire of

Lady Cloud 19,358	16.10

FARMER'S GLORY 5196 (274 J. H. B.)
Son of Grey King 169 J. H. B. and Bonheur 1651 J. H. B.
1st sire of

Maritana 12,039 - -	16.03½
Geneva 13,220 -	15.11
Marie S. 12,043 -	15.06

2d sire of

Baron's Rosette 25,988 -	15.04

FARO 1749.
Son of Vermont 893 and Frankie 17.
2d sire of

Almah of Oakland 11,102	16.14

FASHION 862.
4th sire of

Maple Leaf 4768 -	14.13

FAST BOY 2606.
Son of Bon Ton 1656 and Artless 3992.
1st sire of

Nibbette 11,625	14.07

FAUST 503.
Sire on I. of Jersey, dam Fanny 1343.
4th sire of

Wakena 19,721	16.00

FAYETTE BOY 4737.
Son of Fancy Leo 2728 and Lady Pigot 5798.
1st sire of

Lady Cloud 19,358 -	16.10

FIGARO 154.
Son of Diamond 155 and Flirt 326.
3d sire of

Alice of Salem 5053 -	14.08

FITZ 1988.
Son of Pagan 1800 and Canary Bird 2d 4264.
1st sire of

Bet Arlington 8970	18.11

2d sire of

Countess Queen 13,519	18.03

FLASH 2532.
Son of Pierrot 636 and Belle of Farmington 911.
1st sire of

Rarity 2d 7724	14.02

2d sire of

Lady Cecelia 24,821 -	16.01

FORTUNATUS 1152.
Son of Mogul 532 and Juniata 1289.
1st sire of

Rosebud of Allerton 6352	19.12
Violet of Glencairn 10,221	14.04

2d sire of

Rosebud of Allerton 6352	19.12

FORTUNE 34.
Son of Jack Frost 31 and Violet 23.
4th sire of
Abbie Z. 14,002   -   14.11

FRANK WARREN 1490.
Son of General Warren 1489 and Cowslip 3706.
1st sire of
Negress 7651   14.00
2d sire of
*Fawnette of Woodstock 3710 - 17.08
Enigma 5360 - - - 15.06
Thornedale Belle 5265   14.08
3d sire of
Thornedale Belle 3d 10,459 15.15
Lydia of Libby 11,698   15.03
4th sire of
Belle Thorne 13,369   14.11

FRITZ 565.
Sire on I. of Jersey, dam Winglet 1230.
2d sire of
Dark Cloud 9364 - 15.03½
Rose of Rose Lawn 9365 14.06

GARIBALDI 609.
2d sire of
Empress 6th 3203   17.09¾
3d sire of
Rene Ogden 1568 · 15.00
Belle of Ogden Farm 1570 14.00
4th sire of
Medrena 3939 - - - 17.12
*Matindy 6670   - 16.03
Callie Nan 7959 -   - 16.02
Romp Ogden 3d 5458   - 14.01
Litty 8017   14.00

GEN. BAUM 433.
Son of Challenger 236 and Fairy Belle 1152.
4th sire of
Fair Starlight 1745 -   17.07½

GEN. BEHM 552.
Sire on I. of Jersey, dam Ceres 1327.
3d sire of
Alice Herrick 8787   - 14.14

GEN. GRANT 47.
Son of Gen. Scott 46 and Palestine 26.
2d sire of
Landseer's Fancy 2876 · 22.07½
3d sire of
Perces Lee 5538 - ·- 16.10
Rosabel Hudson 5704 - 15.12
Rosy Dream 9808 - 14.13

GEN. GRANT 1409.
Son of Harry 72 and Funny 3565.
1st sire of
Helen 3556 - - - - 15.09

2d sire of
Meines 3559 - - 14.00
3d sire of
Meines 3d 7741 - - 20.01
Del of Willow Farm 22,464 14.08

GEN. MILES 301.
4th sire of
Baby Buttercup 10,888   14.00

GEN. R. E. LEE 208.
Son of Reefer 209 and Countess 896.
3d sire of
*Dora O. 11,703   - 17.03
4th sire of
*Miskwa 15,472   - 19.06
Rosa of Glenmore 3179   17.10
Embla 4799   17.08

GEN. ROSSER 4189.
Son of Fitz 1988 and Orphan Beauty 3769.
1st sire of
Countess Queen 13,519 - 18.03

GEN. SCOTT 46.
Son of Splendid 2 and Sue 2d 65.
1st sire of
Maggie Mitchell (unreg.)   18.02
Palestine 3d 1104 - 16.04
2d sire of
Palestina 4644 - - 15.05
Lady Fanning 11,169   - 14.06
3d sire of
Landseer's Fancy 2876   22.07½
Attractive Maid 16,925 16.13
Myrtle 2d 211   15.12
4th sire of
Perces Lee 5538   16.10
Canto 7194   - 15.12
Rosabel Hudson 5704 - 15.12
Rosy Dream 9808   14.13
Celia Belle 5865   - 14.03
Myrtle of Ridgewood 7858 14.01

GEN. SHERIDAN 1339.
Son of Admiral 1337 and Loring's Nellie 3424.
3d sire of
Woodland Lass 3444   14.00

GEN. WARREN 1489.
Son of Jerry 15 and Frankie 17.
2d sire of
Negress 7651 -   14.00
3d sire of
*Fawnette of Woodstock 3710 - 17.08
Enigma 5360 -   15.06
Thornedale Belle 5265   - 14.08
4th sire of
Thornedale Belle 3d 10,459 - 15.15
Lydia of Libby 11,698 - 15.03

GEO. WASHINGTON 696.
Son of Sark 123 and Jura 2d 50.
4th sire of

Goldthread 4945	- 16.09
Lady Penn 5314 -	16.00

GERALD 895.
Son of Emperor 287 and Gipsy 319.
1st sire of

Daisy of Clermont 8492	14.00

GIL BLAS 1193.
Son of Mogul 532 and Cybele 136.
1st sire of

Leonice 2d 8842 -	16.08

GILDEROY 2107.
Son of Magnetic 1428 and Jeanne Le Bas 2476.
1st sire of

Chrome Skin 7881	20.10
Lactine 10,680	17.01½
Daisy's Daughter	15.02
Sweet Sixteen 10,682	14.15
Gold Mark 10,727	14.14
Yellow Locust 10,679	14.10½
Mary of Gilderoy 11,219	14.04

GILROY 1653.
Son of Hamilton 1074 and Pet Gilford 3317.
1st sire of

Euphonia 6783	16.00½

GINX 1586.
Son of Ben Rajah 795 and Audrey 1447.
1st sire of

Walkyrie 5708	14.01

GLENGARY 316.
Son of Jupiter 93 and Edna 807
2d sire of

* Katie Kohlman 7270 -	23.10
* Lass Edith 6290	17.00
* Myra 2d 6289	16.00
Trudie 2d 4084	15.00

4th sire of

Renown 13,729	- 14.06

GOLDEN BALL 1474.
Son of Brookside 1104 and St. Catherine 408.
4th sire of

Baby Buttercup 10,888 -	14.00

GOLDEN LION 5239.
Son of Ned Ives 3631 and Hannah Duncan 3d 4029.
1st sire of

Percie 14,937 -	16.12

GOLD PRINCE 2181.
Son of Landseer 331 and Myrtle 2d 211.
1st sire of

Miami Prize 8100	- 14.00

2d sire of

Gold Princess 8809 -	14.12

GOVERNOR 890.
Son of Napoleon 291 and Gracie 769.
2d sire of

Empress 6th 3203 -	- 17.09½
Goddess of Staatsburg 5252	14.08

3d sire of

Pride of Bovina 8050	16.09
Dorothy of Bovina 9878	15.04
Lady Adams 2d 6529	15.03
Miss Baden Baden 14,760	14.14½

4th sire of

Almah of Oakland 11,102	- 16.14

GOV. HAMPTON 2701.
Son of The Hub 1009 and Calla 5th 2215.
2d sire of

Marpetra 10,284 -	14.06

GRAND DUKE 665.
Son of Pilot, Jr., 141 and Duchess 101.
3d sire of

Wakena 19,721 -	- 16.00

GRAND DUKE ALEXIS 1040.
Sire on I. of Jersey, dam Victorine Lachaise 2740.
1st sire of

Chrissy 2d 7720 -	- 16.14
Polynia 10,753	16.07
Corollo 4892 - -	16.02
Countess Croton 5807 -	15.12
*Rosellen 5306	15.05
Roselaine 7167 -	15.01
Cosette 3874 - -	14.10½
Hartwick Belle 7722 -	14.08

2d sire of

Tenella 6712 -	22.01½
Tenella 2d 19,521	18.12
Valhalla 5300 -	17.00
Belle of Patterson 5664 - -	16.06
Gold Trinket 9618	16.02
Rupertina 10,409	15.12½
Azelda 2d 7022 -	15.02
Aldarine 5301 -	15.01½
Alphea Jewel 22,331	14.00

3d sire of

Fadette of Verna 3d 11,122	22.08½
Fairy of Verna 2d 10,973 -	20.03½
Hilda A. 2d 11,120 -	20.00
Gardinier's Ripple 11,698	- 19.12½
Tenella 2d 19,521 -	18.12
Genevieve Sinclair 11,167	16.02
Euphorbia 11,229	14.09½
Signetilia 16,333 -	14.08½
Litty 8017 -	14.00
Sadie's Choice 7979 -	14.00

4th sire of

*Lenoreisa 16,233	21.00
Signetilia 16,333	- 14.08½

GRAY 244 J. H. B.
Son of Duke 76 J. H. B. (Sweepstakes Duke 1905.)
1st sire of

Handsome Myra 14,244	20.08

GRAY COAT 1105.
Son of Yellow Skin 871 and St. Catherine 408.
1st sire of

Lady Louise 4339	-	15.00
Gilt 4th 4208	-	14.00

2d sire of

Sasco Belle 13,601	14.00

GRAY EAGLE 174.
Son of Living Storm 173 and Minnie 398.
4th sire of

Jennie of the Vale 9553 -	14.06½

GRAY FRIAR 567.
Son of Jupiter 92 and Alphea 171.
2d sire of

Lady Alice of Hillcrest 7450	16.03
Pride of Corisande 5323	16.00
Gray Therese 5322	16.00

3d sire of

Princess Mostar 9700	17.03
Lillian Mostar 10,364 -	14.03
Elmora Mostar 15,955	14.00

GRAY KING 169, J. H. B.
3d sire of

Beauty of the Grange 7502 -	23.09
Eveline of Jersey 6781	18.06
*Violet of St. Ouens 8626	17.08

GRAYLOCK 740.
Son of Collamore's Atlantic 739 and Mudine 1864.
1st sire of

Daisy of Belhurst 3114	16.08

3d sire of

Countess Lowndes 26,874	17.08

GREY WETHERSFIELD 1250.
Son of McClellan 6th 181 and Lucy Neal 3d 418.
2d sire of

Maggie 3d 3221	-	17.08

3d sire of

Cowle's Nonesuch 6199	14.12
*Myra Overall 10,317	14.00

GREY OF THE WEST 1317 J. H. B.
Son of Grey King 169 J. H. B.
2d sire of

Beauty of the Grange 7502	23.09
Eveline of Jersey 6781	18.06
*Violet of St. Ouens 8626	17.08

GREY PRINCE 168 J. H. B.
Son of Grey of the West 1317 J. H. B.
1st sire of

Beauty of the Grange 7502	23.09
Eveline of Jersey 6781 -	18.06
*Violet of St. Ouens 8626	17.08

GROTON 2280.
Son of Magnet 968 and Sallie 2640.
1st sire of

Village Maid 7069	14.00

GUY FAWKES 251 J. H.B.
Son of Koffee 233 J. H. B. and Angelica 1738 J. H. B.
1st sire of

Island Star 11,876 -	21.03
Mabel of Trinity 13,694	16.03
Queen of Ashantee 14,554	15.02
Auntybel 15,582	14.09

2d sire of

*Pendule 2d 16,709 - -	16.03
Queen of Ashantee 2d 16,657	14.03½

GUY MANNERING 698
Sire on I. of Jersey, dam Brunette Lass 1780.
1st sire of

Phlox 16,399 -	21.11
Fair Lady 6723	19.00
May Fair 5184	16.07
Cottage Lass 5332	14.08

2d sire of

Fair Lady 6723	-	19.00
May Fair 5184		16.07

GUY WARWICK 1450.
Son of Mercury 432 and Edith 3d 806.
1st sire of

Honey Drop 10,033	- -	14.00½

HADAD 848.
Son of Hannibal 618 and Arango 1597.
1st sire of

Naomi Cramer 8628 -	14.00

HADRIAN 1049.
Son of Bronx Bashan 145 and Dinah 1031.
2d sire of

Nellie Gray of Clermont 10,905	14.01

HALIFAX 681.
Son of Defiance 196 and Hebe 489.
2d sire of

Nordheim Creamer 9758	- -	14.00

HAMILTON 1074.
Son of Marius 760 and Emily Hampton 1912.
1st sire of

Etiquette 4300	- -	15.08
Bellita 4553	-	14.00
Elite 4299		14.00

2d sire of

Euphonia 6783	16.00½

3d sire of

Percie 14,937 -	16.12
Lottie Rex 18,757	14.04

HAMPTON 491.
Son of Black Imperial 255 and Lightfoot 461.
2d sire of

Celia Belle 5865	14.03

HANNIBAL 618.
2d sire of

Vieva 3d 7642	16.05
Avis E. 9714 -	15.14
Naomi Cramer 8628	14.00

3d sire of

Grace Davy 8292	14.02
Baby Buttercup 10,888	14.00

4th sire of

Bessie Ridgeley 8293	14.11½

HAPPY 211 J. H. B.
*Son of Vertumnus 161 J. H. B. and Garenne
24,534.*
1st sire of

St. Jeannaise 15,789	16.04

2d sire of

*Pendule 2d 16,709	16.03

HARPADO 1859.
Son of Knave 1856 and Betta 3075.
1st sire of

Alfleda 6744 -	16.04

2d sire of

Alfritha 13,673	15.03

HARRY 72.
Son of Norman 73 and Bessie 111.
2d sire of

Helen 3556	15.09

3d sire of

Moss Rose of W. F. 5194	18.08½
Lady Love 2d 2212 -	16.08
My Queen 12,614	15.08
Del of Willow F. 22,464	14.08
Meines 3559 -	14.00

4th sire of

Pavon 12,485	14.08
Del of Willow F. 22,464	14.08

HARRY 92.
*Son of Taintor's Bull of '58 306 and Harry's
Dam 258.*
3d sire of

Jo 5th 280	17.08

4th sire of

Ida Bashan 4725	18.00
Vespucia 17,445	14.04

HARTFORD 52.
Son of Pistol 53 and Fanny 72.
2d sire of

Mirtha 3437 - -	17.13½
Mirth's Blanche 19,572	17.13½

3d sire of

Lady Josephine 11,560 (8 days) -	19.02
Merry Burlington 7600 -	15.04

4th sire of

Forsaken 7520	15.01
Coronilla 8367	14.09½

HECTOR 129.
Son of Prince Albert 119 and Victoria 249.
1st sire of

Monmouth Duchess 3895	- 14.07
Countess of Warren 3896	14.00

2d sire of

Warren's Duchess 4622 -	16.01

3d sire of

Dot of Bear Lake 6170	- 19.04
Ida of Bear Lake 6169	16.00
Mary of Bear Lake 6171	14.07

4th sire of

Countess Lowndes 26,874	17.08

HECTOR 260.
Son of Potomac 153 and Promise 751.
4th sire of

Jesse Lee of Labyrinth 5290	14.07

HECTOR 791.
Son of Blucher 48 and Daisy 571.
1st sire of

*Mica 1983 -	15.12
Lilly Cross 13,796	14.03

3d sire of

Hillside Gem 16,640	20.00
Lilly Cross 13,796	14.03

HECTOR 3814.
Son of Reward 190 and Peace 330.
1st sire of

Eureka McHenry 8341 -	14.00

HECTOR OF PLYMOUTH ROCK 886.
3d sire of

Queen of De Soto 12,318 -	14.13

4th sire of

Queen of De Soto 12,318	14.13
Fall Leaf 8587 -	14.08
Adora 18,569 '	14.03

HELIERO 478.
Son of St. Helier 45 and Fanchette 66.
1st sire of

Perces Lee 5538	16.10

HERDSMAN 137.
Son of Pilot 3 and Jenny 287.
3d sire of

Urbana 5597 -	16.00
Bathsheba 2556	14.01

4th sire of

Pattie Mc. 3d 4754 -	- 16.08

HERO 90 J. H. B.
Son of Welcome 172 J. H. B., and Musique
1096 J. H. B.
1st sire of

Daisy of St. Peters 18,175	20.05½
Jenny Le Brocq 9757	14.14
Cocotte 11,958	14.12
Satin Bird 16,880	14.10

HERO 126 J. H. B.
Son of Dick 171 J. H. B. and Cowslip 24
J. H. B.
2d sire of

Belle Dame 2d 22,043	- 15.08
Carlo's Fanny 14,951 -	14.00

HERO 840.
Sire J. P. Cushing's Bull, dam Flora 2018.
2d sire of

Duchess Caroline 3d 6089 -	15.08

HIGHLAND BLADE 2164.
Son of Wethersfield 966 and Chloe Daniels
2668.
1st sire of

Jennie of the Vale 9553	- 14.06½

HILLHURST 1210.
2d sire of

Pansy of Bellewood 8904 -	- 18.00

4th sire of

Miss Baden Baden 14,760	14.14½

HIMAN 78.
Son of Brandywine 64 and Fawn 118.
3d sire of

Magnibel 7976 -	14.12

HINDOO 2282.
Son of Noble 93 J. H. B. and Le Gallais'
Mermaid 4954.
1st sire of

Maudine of Elmwood 8718 -	16.15

HOBART BULL 285.
4th sire of

Beeswax 9807 -	- 17.05

HOCKANUM 792
Son of Blucher 48 and Dewdrop 1158.
1st sire of

Copper 1979	15.07

2d sire of

Mica 1983	15.12

4th sire of

Lady Bidwell 10,308 -	15.12

HOMER H. 3683.
Son of the Squire 1298 and Gilda 2779.
1st sire of

Jenny Dodo H. 14,448	21.08

HOPEWELL 136.
2d sire of

May Blossom 5657 - -	18.11
*Filbert 3680 - -	15.12
Beauty Bismarck 4967 -	14.01

4th sire of

Alhena 15,995 - -	- 16.08

HORNBEAM 2123.
Son of Marius 760 and Emily Hampton
1912.
1st sire of

Troth 6139 -	16.05

HOTSPUR 206.
Son of Gen. R. E. Lee 208 and Duchess 895.
3d sire of

Rosa of Glenmore 3179	17.10
Embla 4799 -	17.08

4th sire of

Naomi Cramer 8628 -	14.00

HOUSATONIC 704.
Son of Cliff 176 and Hebe 1177.
4th sire of

Belle of Vermillion 8798	15.02
Lillian Mostar 10,364	14.08

HUDSON 116.
3d sire of

Spring Leaf 5796	- 14.00

4th sire of

Belle of Vermillion 8798	15.02
Lillian Mostar 10,364	- 14.08

HUGHES 954.
Sire on I. of Jersey, dam Daffle 2522.
1st sire of

Gentle of Glastonbury 4651	- 14.00

2d sire of

Colt's La Biche 6899	17.02½
Lobelia 2d 6650	14.06

3d sire of

Princess Shielia 7297 -	16.04½
Alfleda 6744 -	- 16.04
Susie La Biche 3d 15,171	14.06½
Lilly Cross 13,796 -	14.08
Baby Buttercup 10,888 -	14.00

4th sire of

Lady Bidwell 10,303 -	15.12
Alfritha 13,673 -	- 15.08
Hurrah Pansy 12,153	14.01½

HURD'S IVANHOE 1522.
Son of Zany 551 and Blooming Beauty 2584.
1st sire of

Orphean 4636	15.07

2d sire of

Belmeda 6229 · ·	18.12
Gem of Sassafras 8434	14.03½

HURRAH 2814.
Son of Col. Crockett 1694 and Village Girl 5744.
1st sire of

Value 2d 6844 -	25.02½½
Hurrah Pansy 12,153	14.01½

HYPERION 589.
Sire on I. of Jersey, dam Spotless 1503.
2d sire of

Hazalena's Butterfly 10,123	14.00

IKE FELCH 1292.
Son of Critic 540 and Maid of Judah 2429.
1st sire of

Lily of Burr Oaks 11,001 -	15.13
Grace Felch 8291 - -	15.00
Milkmaid of Burr Oaks 9035	14.05

3d sire of

Wakena 19,721	16.00

INACHUS 928.
Son of Rioter 2d 469 and Dido 1234.
4th sire of

*Pedro Alphea 13,889	15.05

INDIAMAN 2071.
Son of Vespucius 758 and Ibex 2724.
1st sire of

Vespucia 17,455	14.04

2d sire of

Vespucia 17,455	14.04

INTREPID 5511.
Son of Mercury 432 and Clotho 2566.
1st sire of

Fillpail 16,530	15.11

IRON BANK 1120.
Sire on the I. of Jersey, dam Birdie 2611.
1st sire of

*Lebanon Wife 6102	18.09
Willis 2d 4461	16.03
Vaniah 6597 ·	15.09½
Lebanon Daughter 6106	14.04
Lebanon Lass 6108 - -	14.02

2d sire of

Blossie Reynolds 6082	16.03½
Dom Pedro's Julian 8631	16.00
Royal Sister 12,457	14.11
Home Matron 6707	14.00

3d sire of

Royal Sister 12,457 -	14.11

IRON DUKE 18
Son of Pilot 3 and Fairy 10.
3d sire of

Mary M. Allison 6308 -	20.14
Cerita of Meadowbrook 5056 -	17.08
Pattie Mc 3d 4754 - -	16.08
Charmer 4771 ·	14.12
Epigæa 4631 - -	14.07
Queen of Prospect 11,997 -	14.02

4th sire of

Hazen's Nora 4791 - -	20.04
Rosebud of Allerton 6352 -	19.12
*Matindy 6670	16.03
Calypris 5943 -	15.09½
Arawana Buttercup 6052 -	15.05
Arawana Poppy 6053	15.02
Cosetta 15,991 - -	14.11
Belle of Milford 7445 -	14.07
Violet of Glencairn 10,221	14.04
*Lady Golddust 2d 19,861	14.01

ISAAC 42.
3d sire of

Oxalis 606	15.00

4th sire of

Oxalis 2d 15,631	15.00

ISAAC B. 1951.
Son of Matchem 747 and Athena 2152.
1st sire of

Lily of Maple Grove 5079	16.03

2d sire of

Lida Mullin 9198	16.08
Lizzie D. 10,408 - -	14.00

ISHMAEL HURD 1548.
Son of Major Tunxes 1547 and Buck's Kate 3463.
2d sire of

Cowle's Nonesuch 6199	14.12

ISLANDER 561.
Son of John Le Bos 398 and Ida 1441.
2d sire of

Magnibel 7976	14.12

3d sire of

Chrome Skin 7881	20.10
Lactine 10,680 -	17.01½
Daisy's Daughter -	15.02
Sweet Sixteen 10,682	14.15
Gold Mark 10,727 - -	14.14
Yellow Locust 10,679 ·	14.10½
Mary of Gilderoy 11,219	14.04

ISLAND LORD 3322.
Son of Antelope 1927 and Mulberry 2d 2199.
2d sire of

Baby Buttercup 10,888 -	14.00

JACHIN 1220.
Son of Yankee 1008 and Jennie 3d 2244.
2d sire of

Countess Lowndes 26,874 -	17.08

JACK DASHER 932.
Son of Tom Dasher 490 and Judy 691.
1st sire of

Peggy Leah 3097	-	18.12
Duchess of Argyle 3758		14.13

2d sire of

Polly Clover 7052	- -	16.15

JACK FROST 31.
Son of Splendeus 16 and Jessie 28.
3d sire of

Princess of Mansfield 8070	15.02
New London Gipsy 11,667	14.08

4th sire of

Peggy Leah 3097	-	18.12
Pierrot's Lady Bacon 12,482		16.10
Duchess of Argyle 3758		14.13
Chloe Beach 8931		14.08

JACK FROST OF ST. LAMBERT 2419.
Son of Buffer 2055 and Pride of Windsor 483.
1st sire of

Coquette of Glen Rouge 17,559	15.01½
Honeysuckle of St. Annes 18,674	14.14

JACK HORNER 514.
1st sire of

*Lady Caroline 3674	14.14

3d sire of

Mink 2d 3890 -	-	19.11
Oktibbeha Duchess 4422		17.04
Arawana Foppy 6053		15.02
Mink 3d 4868	-	14.09
Adora 18,569 -	-	14.03

4th sire of

Mhoon Lady 6560	-	17.03
Ochra 2d 11,516		16.06½
Julia Evelyn 6007		15.15½
Valerie 6044 -		15.13
Dairy C. 12,227		15.01

JACK LIBBY 3307.
Son of Sweepstakes Duke 1905 and Alpha 2d 5352.
1st sire of

Lydia of Libby 11,698	. -	15.03

JACOB 1377.
Son of Graylock 740 and Miss Jenkins 892.
2d sire of

Countess Lowndes 26,874	-	17.03

JAQUES 63 J. H. B.
Son of Stockwell 2d 24 J. H. B.
1st sire of

Reception 8557	- -	19.03

2d sire of

Dora Neptune 20,318	-	20.00½
Reception 3d 11,025	- ;	15.08½

JASON 1550.
Son of Neptune of the Grange 1549 and Jessica of the Grange 3805.
3d sire of

*Pedro Alphea 13,889	-	15.05

JASON, JR., 3270.
Son of Jason 1550 and Lady Reynolds 3808.
2d sire of

*Pedro Alphea 13,889	15.05

JASON OF DEERFOOT 1636.
Son of Jersey Boy 272 and Jessie 2715.
3d sire of

Percie 14,937	16.12

JASPER 850.
Son of Clement 115 and Canary 327.
2d sire of

Conover's Beauty 12,650	18.00

JAZEL 3501 (159 J. H. B.).
Son of Sweepstakes Duke 1905 (76 J. H. B.) and Flying 18 J. H. B.
1st sire of

Sweetrock 18,256	-	14.11¼
Jazel's Maid 11,011	-	14.08

JENKINS 888.
Sire on I. of Jersey, dam Juliana 2236.
3d sire of

Dorothy of Bovina 9373	15.04

JERRY 15.
2d sire of

Lady Mel 2d 1795		21.00
Couch's Lily 3237		16.09
Lady Love 2d 2212		16.08
Kitty Colt 2213		15.09½
Fragrance 4059	-	14.12
Goddess of Staatsburg 5252		14.08
*Brightness 2211	-	14.00
Actress 2311	-	14.00

3d sire of

Lady Gray of Hill Top 6850		18.12
Belle Grinnelle 4073		18.08
Rosa Miller 4333	-	17.07
Oktibbeha Duchess 4422		17.04
Jersey Cream 3151	-	17.00
Lucky Belle 2d 6087		16.14
Dusky 2525	- -	16.10
Pride of Bovina 8050	-	16.09
*Kitty Lake 8250	-	15.08½
Brightness 3d 14,824		15.05
Olie 4133	- -	15.00
Bloomfield Lady 6912		14.12
Gold Princess 8809		14.12
Aroma 8518	-	14.07
Susette 4068 -		14.04
Rarity 2d 7724 -		14.02
Maggie May 3255		14.02
Creamer 2467		14.01
Pretty 2526	-	14.00
Daisy of Clermont 3492 -	-	14.00
Negress 7651 -		14.00

4th sire of

*Optima 6715 -	23.11
Tenella 6712 - -	22.01½
Croton Maid 5305 -	21.11½
Countess Potoka 7496	18.15
Peggy Leah 3997 - -	18.12
Lady Gray of Hill Top 6850 -	18.12
May Blossom 5657 · -	18.11
Summerline 8001	18.06
Cordelia Baker 8814	17.09
Hepsy 2d 12,008 · - -	17.08
*Fawnette of Woodstock 3710	17.08
Valhalla 5300 - -	17.00
Almah of Oakland 11,102	16.14
Pride of Bovina 8050	16.09
Arawana Queen 5368 -	16.09
Belle of Patterson 5664	16.06
Œnone 8614 -	15.14
Edwina 6713 -	15.13
Valerie 6044 - -	15.13
Fanny Taylor 6714 -	15.12
Princess Belworth 6801	15.10½
Fancy June 6086	15.10
*Kitty Lake 8250	15.08½
Signalana 7719 -	15.04
Dorothy of Bovina 9873	15.04
Usilda 2d 6157	15.02½
Aldarine 5801 -	15.01½
Favorite's Rajah Rex 16,153	15.00
Mary Clover 9998 -	14.15
Miss Baden Baden 14,760	14.14½
Queen of De Soto 12,318 -	14.13
Duchess of Argyle 3758 -	14.13
Louvie 3d 6159 -	14.13
Lady Gray of Hill Top 2d 14,641	14.12
Jersey Cream 2d 8519	14.12
Bell Rex 11,700 -	14.10
Princess Rose 6249 - ·	14.08
Thornedale Belle 5265	14.08
Deborana 4718 -	14.08
Jennie of the Vale 9553 -	14.06½
Maggie May 2d 12,926 -	14.06
Maggie C. 12,216 -	14.06
Jeannie Platt 6005	14.04
Vespucia 17,455 -	14.04
Lottie Rex 18,757	14.04
Prince's Bloom 9729	14.03
Pet Rex 20,166 -	14.02½
Lady Gray of Hill Top 3d 14,642	14.02
Rarity 2d 7724 - -	14.02
Belle Grinnelle 3d 16503	14.02

JERRY 2D 513.
Son of Jerry 15 and Eve 733.
3d sire of

Fancy Juno 6086	15.10
Vespucia 17455	14.04

4th sire of

Judith Coleman 11,391 -	17.05
Lady Cloud 19,358	16.10
Aleph Judea 11,389	15.01¾

JERSEY 9.
Sire on I. of Jersey, dam Daisy 241.
1st sire of

Rose 3d 239	16.00

2d sire of

Cowslip 5th 849	15.04

3d sire of

Copper 1979	15.07

4th sire of

Vixen 7591	17.06
*Pulsatilla 7551	16.03
Rene Ogden 1568 - -	15.00
Abbie Z. 14.02	14.11

JERSEY BOY 92, J. H. B.
Son of Welcome 172 J. H. B. and Sourie
1107 J. H. B.
1st sire of

Oakland's Cora 18,853	19.09½

JERSEY BOY 272.
Son of Czar 272 and Fanny 675.
2d sire of

Christmas Nannie 4075	19.07
Patty of Deerfoot 15,321	16.00
Dena of Deerfoot 15,325	14.08
Daisy of Chenango 18,582	14.07

3d sire of

Polly of Deerfoot 15,328 -	15.00

4th sire of

Countess Queen 13,519	18.03
Percie 14,937 -	16.12

JERSEY GOLDDUST 2134.
Son of Young Yankee 62 J. H. B. and Clelie
2d 64 J. H. B.
2d sire of

*Lady Golddust 2d 19861	14.01

JERSEY HERO 1488.
Son of Bill 50 and Fanny 3703.
3d sire of

*Fawnette of Woodstock 3710 -	17.08
Thornedale Belle 5265	14.08

4th sire of

Thornedale Belle 3d 10,459	15.15
Lydia of Libby 11,698 -	15.03

JERSEY KING 879.
Son of Albert 44 and Grinnella 1302.
1st sire of

*Brightness 3d 14,824	15.05
Susette 4068 · - -	14.04

2d sire of

Countess Potoka 7496	18.15
Prince's Bloom 9729	14.03
Rarity 2d 7724 -	14.02

JERSEY LAD 147.
Son of Comet 86 and Lucy Neal 316.
2d sire of

Pansy of Bellewood 2d 8904	18.00
White Clover Leaf 4512	17.15

4th sire of

Lucy Gaines' Buttercup 5058	14.00

JERSEY PRINCE 1062.
Sire on I. of Jersey, dam Hattie 739.
3d sire of

Celia Belle 5865	14.03

JESSE OF MOBILE 1038.
Sire on I. of Jersey, dam Blossom 2728.
3d sire of

Roland's Bonnie 2d 18,054 -	19.02

JESTER 456.
Son of Santa Ana 221 and Jessie 544.
2d sire of

Maud Lee 2416	23.00

4th sire of

Milkmaid of Burr Oaks 9035	14.05

JEWELER 1385.
Son of Mogul 532 and Jewel Beauty 2d 1701.
1st sire of

Lady Alice of Hillcrest 7450	16.03

3d sire of

Thorndale Belle 3d 10,459 -	15.15
Mitten 13,368	15.11
Alfritha 13,673	15.03
Belle Thorne 13,369	14.11

JIMMY 150, J. H. B.
Son of Sweepstakes Duke 1905 (76 J. H. B.) and Flora 813 J. H. B.
2d sire of

Sweet Sixteen 10,682	14.15

JO BRADLEY 4640.
Son of Pequaboc Chief 2662 and Village Girl 5744.
2d sire of

*Wabash Girl 14,550-	16.00

JOHN ALLEN 1494.
Son of Tom Dasher 420 and Allen's Fawnette 3722.
2d sire of

Cowle's Nonesuch 6199 - '	14.12

JOHN BROWN 67.
Son of Prince John 22 and Victoria 104.
2d sire of

Lady Brown 438	14.00

3d sire of

Silver Rose 4753 -	16.14
Champion Chloe 12,255 -	15.05½
Arietta 5115 - -	15.00
Lady Brown 4th 6911 -	14.12
Lucy Gaines' Buttercup 5058	14.00

4th sire of

Sultana 2d 11,798 -	15.04
Bloomfield Lady 6912 - -	14.12
*Woodland Margaret 6215 -	14.10½
Opaline 7590 -	14.10
Negress 7651	14.00

JOHN BULL 167.
3d sire of

Avis E. 9714 - - -	15.14

4th sire of

Belle of Vermillion 8798	15.02
Oxalis 606 - -	15.00

JOHN BULL 358.
2d sire of

Maid of Amboy 2929 - -	16.01	
Phyllis of Hillcrest 9067	14.12	

3d sire of

Maid of Amboy 2929 -	16.01
Lerna 8684 -	15.12
Iola 4627 - -	15.02½
Phyllis of Hillcrest 9067	14.12
Roll of Honor 13,610	14.12

4th sire of

Hazen's Bess 7429	24.11
Miss Willie Jones 6918	16.04 ·
Zalma 8778	15.05
Faustine 10,354	14.14½
Ideal 11,842 -	14.12½
Roll of Honor 13,610	14.12
Pet Rex 20,166	14.02½
Leruella 22,822	14.01¼

JOHN KNOX 3289.
Son of Red Cloud 2d 2260 and Lady Pigot 2d 5798.
1st sire of

Judith Coleman 11,391 - -	17.05	
Aleph Judea 11,889	15.01¾	

JOHN LE BAS 398 (71 J. H. B.).
Sire on I. of Jersey, dam Patricia 189 J. H. B.
1st sire of

Jeanne Le Bas 2476	15.08

2d sire of

Topsey Roxbury 7796	15.00
Audrey 1447 - - -	14.00
Woodland Lass 3444	14.00

3d sire of

Chrome Skin 7881 -	20.10
*Mollie Garfield 12,172 -	18.07
Panatella 4778	18.03
Oclera 2d 11,516	16.06½
Callie Nan 7959 -	16.02
Merry Burlington 7600 -	15.04
Sweet Sixteen 10,682	14.15
Gold Mark 10,727	14.14
Magnibel 7976	14.12
Cosetta 15,991 - - -	14.11
Yellow Locust 10,679	14.10½
Mary of Gilderoy 11,219	14.04
Clover Top 9910 -	14.00
Village Maid 7069	14.00

4th sire of

Chrome Skin 7881 -	20.10	
Lactine 10,680 - -	17.01½	
Daisy's Daughter (?) -	15.02	
Sweet Sixteen 10,682 -	14.15	
Gold Mark 10,727 -	14.14	
Yellow Locust 10,679 -	14.10½	
Lottie Rex 18,757 -	14.04	
Mary of Gilderoy 11,219	14.04	
Walkyrie 5708	13.01	

JOHN RIDGELY 3045.

Son of Young Sir Davy 3034 and Button 2d 3160.

1st sire of

Bessie Ridgely 8293	14.11½

JOHN STREET 6156.

Son of Double Prize 3117 and Flirt of Ipswich 6607.

1st sire of

Countess of Scarsdale 18,633	14.06

JOHN T. NORTON 177.

Sire on I. of Jersey, dam Lily 1.

3d sire of

Lady Louise 4339	15.00
Gilt 4th 4208	14.00

4th sire of

Estrella 2831 -	14.12
Gilt Edge 2d 4420	14.00
Sasco Belle 18,601 -	14.00

JOSEPH 3419.

Sire on I. of Jersey, dam Josephine 1621 J. H. B.

2d sire of

Grinnell Lass 11,859 -	16.10

JOSEPH L. 148.

Son of Jerry 15 and Gipsy 319.

2d sire of

Gold Princess 8809	14.12

JOVE 179.

Sire on I. of Jersey, dam St. Catherine 408.

2d sire of

Lady Louise 4339	15.00

3d sire of

Cenie Wallace 2d 6557	15.04½
Bettie Dixon 4527	15.00
Sasco Belle 18,601	14.00

4th sire of

Yellow Locust 10,679	14.10½

JOVE, JR., 870.

Son of Jove 179 and Hebe 4th 1180.

2d sire of

Cenie Wallace 2d 6557	15.04½
Bettie Dixon 4527	15.00

4th sire of

Marpetra 10,284	14.06

JUNIUS 204.

Son of Iron Duke 18 and Juno 2d 515.

1st sire of

Bathsheba 2556	'14.01

3d sire of

Cosetta 15,991	14.11
Lady Golddust 2d 19,861	14.01

4th sire of

Dom Pedro's Julian 8631 -	16.00

JUPITER 93.

Son of Saturn 94 and Rhea 166.

1st sire of

Europa 176 -	15.00

2d sire of

Eurotas 2454 -	22.07
*Locusta 5143	21.07
Phædra 2561	19.13
Nymphæa 5141	18.07½
Torfrida 3596	17.06½
Lass Edith 6290	17.00
Euphrates 9778	16.05
Lerna 3634	15.12
Zalma 8778 -	15.05
Purest 13,730 -	15.04
Clytemnestra 2455 -	15.03½
Reality 16,537 -	15.03½
Iola 4627 -	15.02½
Marvel 13,734 -	15.01
Nimble 22,335	14.10
Smoky 13,733 -	14.09
Ideal Alphea 18,755 -	14.06
Renown 13,729	14.06
Richness 16,536 -	14.06
*Leda 799 -	14.05½
Alphea Star 16,532 -	14.04½
Alphetta 16,531	14.02½
Vestina 2458 -.	14.02
Lernella 22,322 -	14.01¼
Ballet Girl 18,750	14.01
Alphea Jewel 22,331 -	14.00
Edith 4th 817 -	14.00

3d sire of

*Katie Kohlman 7270 -	23.10
Phædra 2561	19.13
Nymphæa 5141	18.07½
Colie 8309	18.04
Zampa 2194 -	18.00
Zittella 2d 11,922 -	17.06½
*Blossom of Hanover 13,655	17.08
*Lanice 13,656 - - -	17.08
*Lass Edith 6290 -	17.00
Typha 5870, -	16.11
Lady Alice of Hillcrest 7450	16.03
Clytemnestra 2d 5868 -	16.00
Gray Therese 5322 -	16.00
Pride of Corisande 5323	16.00
*Myra 2d 6289	16.00
Bessie S. 5002	16.00
Fillpail 16,530 -	15.11
Malope 2d 11,923	15.10
Idalene 11,841	15.08½
Crust 4775	15.07
Purest 13,730	15.04
Reality 16,537 -	15.03½
Clytemnestra 2455	15.03½
Trudie 2d 4084 -	15.00
Hebe 3d 3613	15.00
Faustine 10,354	14.14½
Ideal 11,842	14.12½
Estrella 2831 - -	14.12
Hartwick Belle 7722	14.08
Richness 16,536	14.06
Renown 13,729	14.06
*Leda 799	14.05½
Kate Daisy 8204	14.04
Alphetta 16,531 -	14.02½
Bessie Bradford 7269	14.02
Lernella 22,322 -	14.01½
Robinette 7114 -	14.01
Honey Drop 10,033	14.00½
Rioter 2d's Venice 3658	14.00
St. Nick's Flora 16,195	14.00

4th sire of

*Katie Kohlman 7270	23.10
Little Torment 15,561	23.02½
Bomba 10,330 -	21.11½
Valma Hoffman 4500	21.09
Nymphæa 5141	18.07½
Oak Leaf 4769	17.10
Zitella 11,922 - -	17.08½
*Blossom of Hanover 13,655	17.08
*Lanice 13,656 - -	17.08
Princess Mostar 9700	17.03
*Lass Edith 6290 -	17.00
Miss Willie Jones 6918 -	16.04
Lily of Maple Grove 5079 -	16.03
Corn 10,504	16.02
*Myra 2d 6289 - - -	16.00
Clytemnestra 2d 5868 -	16.00
Malope 2d 11,923 - -	15.10
Niva 7523- -	15.08
Purest 13,730 -	15.04
Nazli 10,827 - -	15.03½
Bessie Bradford 2d 7271 -	15.02
Forsaken 7520 -	15.01
Marvel 18,734	15.01
Lady Louise 4339	15.00
Maple Leaf 4768 -	14.12
Smoky 13,733 - -	14.09
Hartwick Belle 7722 - -	14.08
Ideal Alphea 18,755	14.06
Richness 16,536 -	14.06
Alphea Star 16,552	14.04½
Lillian Mostar 10,864 - -	14.03
*Lady Golddust 2d 19,861	14.01
Robinette 7114 - -	14.01
Gilt 4th 4208 - -	14.00
Spring Leaf 5796 - -	14.00
St. Nick's Flora 16,195 -	14.00
Alphea Jewel 22,331	14.00
Elmora Mostar - -	14.00
Gilt Edge 2d 4420 - - -	14.00

JUPITER 122.

Son of Sark 123 and Jura 224.

2d sire of

Jo 5th 280	17.08

3d sire of

Ida Bashan 4725	18.00

4th sire of

Lady Alice of Hillcrest 7450	16.03
Gazella 3d 9355	16.03
Lady Penn 5314 -	16.00
Gray Therese 5322 -	16.00
Oitz 8649 - - -	15.01

JUPITER 983.

Son of Gray Eagle 174 and Clover 2580.

3d sire of

Jennie of the Vale 9553 - -	14.06½

KAGO 1353.

Son of Sam Weller, Jr., 1352 and Victoria 3175.

1st sire of

Masena 25,732 - -	20.07

3d sire of

La Pera 2d 13,404 - -	14.08

KALULA 2859.

Son of Lord Byron 707 and Lady Gwendoline 2873.

1st sire of

La Pera 2d 13,404	14.08

KAULBACH 185.

Son of Uncle Pete Noz 186 and Katy Darling 2d 435.

2d sire of

Etiquette 4300	15.08
Elite 4299	14.00

3d sire of

Rose of Hillside 3866 -	14.03½

4th sire of

Countess Lowndes 26,874 - -	17.08

KEARSARGE 257.

Son of Bashan 146 and Lady Webster 638.

2d sire of

Miss Willie Jones 6918	16.04

3d sire of

Pyrrha 6100 -	16.14½

4th sire of

Hazen's Bess 7329 - -	24.11
Gardinier's Ripple 11,693 -	19.12½
Gold Thread 4945 - - -	16.09
Nimble 22,335	14.10

KHEDIVE 103 J. H. B.

Son of Leo 198 J. H. B. and Coomassie 11,874.

1st sire of

Princess 2d 8046	46.12½
Ona 7840 -	20.18
Blonde 2d 9268	14.04
Daisy Queen 9619	14.00

2d sire of

Oxford Kate 13,646 -	39.12
Little Torment 15581	23.02½
Daisy Brown 12213 - -	17.06
Princess of Ashantee 13,487	16.05
St. Jeannaise 15,789 - -	16.04
Odelle Sales 15,564 - - -	16.03
Desire 9654 - -	16.03
Rose of Oxford 13,469 - -	15.14½
Romping Lass 11,021 - -	15.00
Ada Minka 15,562 - -	14.00

KHEDIVE 1051.

Son of Nye 667 and Fairy 2d 1412.

3d sire of

Atlanta's Beauty 12,949 -	21.03

KING 238 J. H. B.

1st sire of

King's Trust 18,946 -	18.00

KING HAROLD 344.

Sire on I. of Jersey, dam Mabille 907.

1st sire of

Belle of Middlefield 1516 - -	16.03

2d sire of

Princess Bellworth 6801 -	15.10½
Chloe Beach 3931	14.08

3d sire of

Arawana Buttercup 6052	15.05
Princess Rose 6249 -	14.08

4th sire of

Rosy Kate's Rex 13,192	18.08
Maggie Rex 28,623	17.00½

KING OF FAIRVIEW 778.
Son of Rob Roy 17 and Eugenie 792.

1st sire of

Fair Starlight 1745 - -	17.07½
Jersey Cream 2d 8519 -	14.12

2d sire of

Katie Bashford 15,982 - -	17.00

3d sire of

Hillside Gem 16,640 - -	20.00
Olie's Lady Teazle 12,307 -	16.05

KING PHILIP 335.
Son of Comet 130 and Jura 224.

2d sire of

Audrey 1447 - -	14.00

4th sire of

Walkyrie 5708 - -	14.01

KING PIN 1878.
Son of Mercury 432 and Edna 3d 668.

1st sire of

Bessie Bradford 7269	14.02

2d sire of

Bessie Bradford 2d 7271 -	15.02

KNAVE 1856.
Son of Hughes 954 and Dusky 2525.

1st sire of

Colt's La Biche 6399	17.02½

2d sire of

Alfleda 6744 -	16.04
Susie La Biche 3d 15,171	14.06½

3d sire of

Alfritha 18,673	15.03

KOBA 416.
Son of Bismarck 292 and Dahlia 401.

2d sire of

Œnone 8614 -	15.14

3d sire of

Thisbe 2d 19,521	18.12

KOFFEE 233 J. H. B.
Son of Nonpareil 37 J. H. B. and Coomassie 11,874.

1st sire of

Young Garenne 13,641	17.08
Gazelle 15,961	14.00

2d sire of

Island Star 11,876	21.08

Mabel of Trinity 18,694	16.03
Queen of Ashantee 14,554 -	15.02
Auntybel 12,582	14.09

3d sire of

*Pendule 2d 16,709 - -	16.03
Queen of Ashantee 2d 16,657	14.03½

LANCASTER 149.
Sire on I. of Jersey, dam Bluebell 116.

2d sire of

Epigæa 4631	14.07

LANDSEER 331.
Sire on I. of Jersey, dam Dazzle 379.

1st sire of

Landseer's Fancy 2876	22.07½

2d sire of

Julia Walker 10,133 -	15.12
Rosabel Hudson 5704 - -	15.12
Rosy Dream 9808 - -	14.13
Little Sister 11,666 -	14.12
Queen Fannie 10,275 -	14.02
Miami Prize 8100 -	14.00

3d sire of

Pride of Eastwood 10,227 -	20.13
*Mary Walker 11,303 -	18.12
Pierrot's Picture 12,481 -	16.00
Pierrot's Lady Hayes 11,672 -	15.12
Lady Hayes 10,136 -	15.12
Gold Princess 8809 - -	14.12
Pierrot's Countess 12,480 - -	14.00

4th sire of

Lady Hayes 10,136	15.12

LANDSEER 2d 788.
Son of Landseer 331 and Sylph 615.

2d sire of

*Mary Walker 11,303 -	18.12

3d sire of

Lady Hayes 10,136	15.12

LAVAL 506.
Son of Defiance 196 and Lisette 492.

1st sire of

Lily of St. L. 5120	14.00

2d sire of

Sweetbriar of St. L. 5481 -	22.12
Jolie of St. L. 5126 -	15.13½
Cupid of Lee Farm 5997	14.06
Pearl of St. L. 5527	14.02

3d sire of

Honeymoon of St. L. 11,221	20.05½
Coquette of Glen Rouge 17,559	15.01½

4th sire of

Mermaid of St. L. 9771	25.13½

LAWRENCE 61 (84 J. H. B.).

1st sire of

Memento 1913	14.05
Turquoise 1129	14.08

2d sire of

Lady of Belle Vue 7705	15.11

Countess Gasela 9571	15.11
Witch Hazel 4th 6131	15.05½
Ma Belle 4942	15.00
Fall Leaf 8587	14.08
Lorella 12,913 -	14.07
Allie Minka 2982 -	14.06½
Irene of Short Hills 5137	14.06½
Cigarette 2849	14.04
Muezzin 3670	14.00

3d sire of

Bonnie Yost 7943 -	18.02
Rosy Dream 9808	14.13
Ada Minka 15,562	14.00

4th sire of

Attractive Maid 16,925	16.13

LE BROCQ'S PRIZE 3350.

Sire on I. of Jersey, dam Matin 1629 J. H. B.

1st sire of

Viva Le Brocq 13,702	17.07
Eclipse 14,427	15.12
Prize Rose 16,309	15.01
Medrie Le Brocq 8888	14.07
Elinor Wells 12068	14.00
La Rosa 10,078 - -	14.00
Birdie Le Brocq 17263	14.00

LE GELE 2694.

Sire on I. of Jersey, dam Fairy 2d 4594.

2d sire of

Countess Coomassie 19,339	15.08½

LEMON 170 J. H. B.

Son of Young Glory 137 J. H. B. and Pale Topsey 563 J. H. B.

1st sire of

Nelly 6456 -	21.00

2d sire of

Nancy Lee 7618 -	26.08½
Daisy of St. Peter's 18,175	20.05½
Miss Vermont 7698 -	16.05
Brenda 3025 J. H. B.	14.00

LEONIDAS 3010.

Son of Signal 1170 and Geranium 3963.

2d sire of

Atlanta's Beauty 12,949	21.03

L'EMPEREUR 461.

Sire on I. of Jersey, dam Lively 1167.

4th sire of

Naomi's Pride 16,745	15.02

LEO 198 J. H. B.

Son of Welcome 166 J. H. B. and Silver Star 215 J. H. B.

2d sire of

Princess 2d 8046 -	46.12½
Ona 7840 -	20.13
Blonde 2d 9268 -	14.04
Ballet Girl 18,750	14.01
Daisy Queen 9619 -	14.00

3d sire of

Oxford Kate 13,646 -	39.12
Little Torment 15,581	23.02½
Daisy Brown 12,213 - -	17.06
Princess of Ashantee 13,467	16.05
St. Jeannaise 15,789 - -	16.04
Odelle Sales 15,564	16.03
Desire 9654 - - -	16.03
Rose of Oxford 13,469	15.14½
Romping Lass 11,021	15.00
Ada Minka 15,562	14.00

4th sire of

Elsie Lane 13,302 -	15.12

LINDO 233.

Son of Jupiter 122 and Carita 730.

3d sire of

Lady Penn 5314 -	16.00

4th sire of

Mirth's Blanche 19,572	17.13½

LITCHFIELD 674.

Son of Hopewell 186 and La Belle Helene 457.

1st sire of

May Blossom 5657	18.11
*Filbert 3630 -	15.12
Beauty Bismarck 4967	14.01

3d sire of

Alhena 15995	16.03

LITTLE JOKER 693.

Son of Dolphin 2d 468 and Sylphide 169.

2d sire of

Typha 5870	16.11

LIVING STORM 173.

Son of McClellan 25 and Pansy 8.

1st sire of

Allen's Fawnette 3722	14.00

2d sire of

Peggy Leah 3097	18.12
Dimple 3248 -	16.11
Cascadilla 3103	15.12
*Kitty Lake 8250 -	15.08½
*Churchill's Betsy 4105	14.00

3d sire of

May Blossom 5657	18.11
Silveretta 6852 - -	16.09
Princess Shiela 7297	16.04½
Tobira 8400 - -	15.13
Champion Chloe 12,255	15.05½
Dairy C. 12,227 -	15.01
Mary Clover 9998 -	14.15
Coronilla 8867	14.09½
Deborana 4718	14.08
Maggie C. 12,216	14.06
Maggie May 2d 12,926	14.06
Gilt Edge C. 12,223	14.03½
Minnie Lee 2d 12,941	14.03
Therese M. 8364 -	14.02
Webster's Pet 4103 -	14.02
Jessie Leavenworth 8248	14.00

4th sire of

Hazen's Nora 4791	20.04
Cordelia Baker 8814	17.09
Polly Clover 7052	16.15
Silveretta 6852 -	16.09
Princess Shiela 7297	16.04½
Albena 15,995	16.03
Tobira 8400 -	15.13
Orphean 4636 - -	15.07
Champion Chloe 12,255	15.05½
Dairy C. 12,227 -	15.01
Mary Clover 9998 -	14.15
Cowle's Nonesuch 6199 -	14.12
Lady Gray of Hill Top 14,641	14.12
Bell Rex 11,700	14.10
Coronilla 8367	14.09½
Maggie C. 12,216	14.06
Maggie May 2d 12,926	14.06
Gilt Edge C. 12,223	24.03½
Minnie Lee 2d 12,941	14.03
Webster's Pet 4108	14.02
Therese M. 8364 - -	14.02
Lady Gray of Hill Top 3d 14,642	14.02
Hurrah Pansy 12,153	14.01½
Baby Buttercup 10,888 -	14.00
Jessie Leavenworth 8248	14.00

LONGFELLOW 818.
Son of Tancred 501 and Undine of Oyster
Bay 1738.
1st sire of

Bessie S. 5002	16.00

2d sire of

Belle of Vermillion 8798	15.02
Lillian Mostar 10,364	14.03

LOOKOUT 121 J. H. B.
3d sire of

Lady Alice of Hillcrest 7450	16.03

LOPEZ 313.
Sire on I. of Jersey, dam Amy 395.
1st sire of

Bertha Morgan 4770	19.06
Mollie Brown 7831	16.00

2d sire of

Lydia Darrach 4903	17.14
Violet of Glencairn 10,221	14.04

3d sire of

Orphean 4636 -	15.07

4th sire of

Niobe of St. L. 12,969	21.09½
Belmeda 6229 - -	18.12
Olie's Lady Teazle 12,307 -	16.05
Gem of Sassafras 8484	14.08½

LORD ANGLESEA 4537.
Son of Apollo 108 J. H. B. and Brunette Le
Gros 9755.
1st sire of

Belle Grinnelle 3d 16,503 - -	14.02

LORD AYLMER 1067.
Son of Lord Lisgar 1066 and Pauline 494.
1st sire of

Melia Ann 5444	18.00½

LORD BALTIMORE 743.
Son of Fairfax 530 and Fides 51.
2d sire of

Maudine of Elmwood 8718	16.15

LORD BRONX 938.
Son of Bully Bronx 604 and Sukey 2d
1224.
2d sire of

Hazen's Bess 7329	24.11

LORD BRONX 2D 1730.
Son of Lord Bronx 938 and Picture 1533.
1st sire of

Hazen's Bess 7329	24.11

LORD BYRON 707.
Sire on I. of Jersey, dam Black Bess 1788.
2d sire of

La Pera 2d 13,404	14.08

LORD DERBY 203.
Son of Stalwart 265 and Florien 660.
3d sire of

Nelida 2d 8227 -	15.02½

4th sire of

Lady Josephine 11,560 (8 days)	19.02
Lutea 4563 -	18.03
*Kalmia 4561 - -	15.00

LORD FRANCIS 1857.
Son of Pierrot 636 and Fanny 72.
1st sire of

Canto 7194	15.12

LORD LAWRENCE 1414.
Son of Lawrence 61 and Lady Mary 1148.
1st sire of

Lady of Belle Vue 7705 -	15.11
Countess Gasela 9571	15.11
Witch Hazel 4th 6131	15.05½
Fall Leaf 8587 -	14.08
Lorella 12,913	14.07

2d sire of

Rosy Dream 9808	14.13

LORD LISGAR 1066.
Son of Victor Hugo 197 and Pauline 494.
1st sire of

Sweet Briar of St. L. 5481	22.12
Jolie of St. L. 5126 -	15.13½
Duchess of St. L. 5111 -	15.13
Clematis of St. L. 5478 - -	14.03

2d sire of

Ida of St. L. 24,990 -	30.02½
Mermaid of St. L. 9771	25.13½
*Allie of St. L. 24,991 -	24.00
Honeymoon of St. L. 11,221	20.05½
*Variella 6387 -	18.03½
Melia Ann 5444 -	18.00½
Cowslip of St. L. 8349 -	17.12
*Brenda of Elmhurst 10,762	17.04½

Minette of St. L. 9774 — 17.04
Chamomilla 7552 - — 16.10
Diana of St. L. 6636 — 16.08
*Pulsatilla 7551 - — 16.03
Carrie Pogis 22,568 - — 15.09
May Day Stoke Pogis 28,353 — 15.03
Coquette of Glen Rouge 17559 - — 15.01½
Honeysuckle of St. Annes 18,674 — 14.14
Bonnie 2d 5742 - — 14.11½
Uinta 5743 - — 14.10
Nora of St. L. 12,962 - — 14.07
Jessie Brown of Maxwell 7266 — 14.07
Moss Rose of St. L. 5114 — 14.00½

3d sire of

Mary Anne of St. L. 9770 — 36.12½
Naiad of St. L.12,965 — 22.02½
Niobe of St. L. 12,969 — 21.09½
Rose of St. L. 20,426 - — 21.03½
Rioter Pink of Berlin 23,665 — 19.14
Crocus of St. L. 8351 — 17.12
Judith Coleman 11,391 - — 17.05
*Dido Miss 8759 — 17.01
Carrie Pogis 25,568 — 15.09
Maggie Sheldon 28,583 — 15.03
Aleph Judea 11,389 — 15.01¾
Rioter's Nora 14,882 — 14.12
Rioter's Beauty 14,894 - — 14.00

4th sire of

Rioter's Beauty 14,894 — 14.00

LORD LONSDALE 305.

Son of Premium 7 and Duchess 2.

2d sire of

St. Nick's Flora 16,195 — 14.00

4th sire of

Cascadilla 3108 — 15.12
Mary Clover 9998 — 14.15

LORD MONCK 304.

Son of Victor Hugo 197 and Pride of Windsor 483.

2d sire of

Pearl of St. L. 5527 — 14.02
Moss Rose of St. L. 5114 — 14.00½

3d sire of

Mermaid of St. L., 9771 — 25.13½
Naiad of St. L. 12,965 - — 22.02½
Rioter Pink of Berlin 23,665 — 19.14
Crocus of St. L. 8351 — 17.12
Judith Coleman 11,391 — 17.05
*Dido Miss 8759 — 17.01
Moth of St. L. 9775 — 16.02
Aleph Judea 11,389 - — 15.01¾
Coquette of Glen Rouge 17,559 — 15.01½
Honeysuckle of St. Annes 18,674 — 14.14

4th sire of

Rose of St. Lambert 20,426 — 21.03½
Rioter's Ruth 14,882 - — 14.12
Rioter's Beauty 14,894 - — 14.00

LORD NELSON 860.

Son of Derby 253 and Knapp Cow 2172.

2d sire of

Oak Leaf 4769 — 17.10
Maple Leaf 4768 - — 14.12
Spring Leaf 5796 — 14.00

3d sire of

Gardinier's Ripple 11,693 — 19.12½
Countess Lowndes 26,874 — 17.08
Euphorbia 11,229 - — 14.09½

LORD OGDEN 69.

Son of Adonis 89 and Big Duchess 58.

2d sire of

Canto 7194 - — 15.12
Lady Ives 3d 6740 — 14.08
Fandango 12,908 - — 14.03

3d sire of

Alfleda 6744 - — 16.04
Myra Overall 10,317 - — 14.00

4th sire of

Alfritha 13,673 — 15.03

LORNE 5248.

Son of Lord Lisgar 1066 and Favorite of St. Lambert 5118.

2d sire of

Maggie Sheldon 23,583 — 15.03

LOYAL 108 J. H. B.

Son of Yankee 1003 (27 J. H. B.) and Stella 705 J. H. B.

2d sire of

*Mabel of St. Marys 8627 - — 18.03

3d sire of

Viva Le Brocq 13,702 - — 17.07
Belle Grinnelle 3d 16,503 — 14.02

LOYAL SATURDAY 215 J. H. B.

1st sire of

Highfield Lass 22,036 — 14.01

LYMAN 793

Son of Hockanum 792 and Ann 1974.

3d sire of

Lady Bidwell 10,303 — 15.12

MACGREGOR 2178.

Son of Rob Roy 17 and Pansy 7th 130.

1st sire of

Lucy Gaines' Buttercup 5058 — 14.00

MACK 722.

Son of Clement 115 and Sunflower 351.

1st sire of

Buttery 3502 - - — 14.01

2d sire of

Alice of the Meadows 20,748 — 14.12
Litza 6338 - — 14.03
Pixie 4115 — 14.00

3d sire of

Bonnie Yost 7943 — 18.02
Leoni 11,868 - — 18.02
Dudu of Linwood 8336 — 16.15
Lucetta 6856 - — 14.03
*Myra Overall 10,317 — 14.00

4th sire of

Armon 10,862 -	16.13½
Dora Doon 12,909	15.00

MAGNET 968.
Son of Mr. Micawber 556 and Mabel 2544.

1st sire of

Woodland Lass 3444	14.00

2d sire of

Mink 2d 3890	19.11
Mink 3d 4868 -	14.09
Village Maid 7069	14.00

3d sire of

Mhoon Lady 6560	17.03
Julia Evelyn 6007	15.15½

4th sire of

*Wabash Girl 14,550	16.00
Therese M. 8864	14.02

MAGNETIC 1428.
Son of Islander 561 and Azalea 1443.

1st sire of

Magnibel 7976	14.12

2d sire of

Chrome Skin 7881	20.10
Lactine 10,680	17.01½
Daisy's Daughter - - -	15.02
Sweet Sixteen 10,682	14.15
Gold Mark 10,727	14.14
Yellow Locust 10,679	14.10½
Mary of Gilderoy 11,219	14.04

MAJESTIC 152.
Sire on I. of Jersey, dam Clio 45.

3d sire of

Mary M. Allison 6308	20.14
*Miskwa 15.472	19.06
*Dora O. 11,703	17.03

MAJESTIC 2d 1201.
Son of Majestic 152 and Lurline 1325.

2d sire of

Mary M. Allison 6308	20.14
*Miskwa 15,472	19.06
*Dora O. 11,703	17.03

MAJOR 75.
Son of Colonel 76 and Countess 114.

3d sire of

Countess of Lakeside 12,135	19.07
Judy 691 - -	19.00
Countess Micawber 1759	16.08
Lucy Gray 2746	15.13
Sylvia 687	15.08
Jersey 3260	15.06
Topsy of Roxbury 7796	15.00
Angela 1682 -	14.02

4th sire of

Jenny Dodo H. 14,448 -	21.08
Miss Blanche 2515 (10 days)	20.09
Duchess of Bloomfield 3653	20.00½
Countess of Lakeside 12,135	19.07
Ebon Edith 10,653	19.01
Su Lu 4705	17.15

Vixen 7591 -	17.06
Pattie Mc. 3d 4764	16.08
*Pulsatilla 7551	16.03
Helen 3556	15.09
Letitia 3977 -	15.05
Arawana Poppy 6053	15.02
Tilda 3720 -	15.00
Gilda 2779	14.06
Bathsheba 2556	14.01

MAJOR 378.
Son of Comet 223 and Duchess 550.

2d sire of

Topsey K. 22,769	14.00

3d sire of

Miss Blanche 2515 (10 days)	20.09

4th sire of

Roland's Bonnie 2d 18,054	19.02

MAJOR 883.
Son of Lord Lonsdale 305 and Major's dam 2227.

3d sire of

Cascadilla 3103	15.12

MAJOR ADAMS 1044.
Son of Duke of Wellington 35 and Lavina 1079.

1st sire of

Lucy Gray 2746	15.13

2d sire of

Jenny Dodo H. 14,448	21.08

MAJOR OF STAATSBURG 679.
Son of Napoleon 291 and Minnie 771.

2d sire of

Celia Belle 5865 -	14.03

MAJOR TUNXES 1547.
Son of Sir John 525 and Buck's Lop Horn 3462.

3d sire of

Cowle's Nonesuch 6199	14.12

MALCOLM 71.
Son of Harry 72 and Brenda 2d 107.

2d sire of

Lady Love 2d 2212	16.08
My Queen 12,614	15.08

MANDARIN 1041.
Son of Orange Skin 19 J. H. B. and Fille de l'Air 2474.

1st sire of

Panatella 4778 -	18.03

2d sire of

Cosetta 15,991	14.11

MANFRED 510.
Sire on I. of Jersey, dam Mildred 1335.

3d sire of

Alfleda 6744 - -	16.04
Nannie Fitch 9143	14.04

4th sire of

Alfritha 13,673	15.03

MARIUS 760.

Son of Willie Boy 434 and Lady Mary 1148.

1st sire of

Welma 5942	17.08
Chenda 4599 -	15.09½
Calypris 5943	15.04½
Bonmari 7019	14.00

2d sire of

*Optima 6715	23.11
Tenella 6712 -	22.01½
Croton Maid 5305	21.11½
Vixen 7591 -	17.06
Beeswax 9807	17.05
Valhalla 5300 -	17.00
Belle of Patterson 5664	16.06
Troth 6130 -	16.05
Busy Bee 6336	16.04
*Matindy 6670	16.03
Œnone 8614	15.14
Edwina 6713	15.13
Fanny Taylor 6714	15.12
Denise 8281 - -	15.09
Etiquette 4300 -	15.08
Signalana 7719 - -	15.04
Jewel 3d —— (A. H. B.)	15.04
Aldarine 5301	15.01½
Dora Doon 12,909 -	15.00
*Queen of Maple Dale F. 7036	14.14
Rosy Dream 9808 -	14.13
*Woodland Margaret 6215	14.10½
Opaline 7590 -	14.10
Medrie Le Brocq 8888	14.07
Marpetra 10,284	14.06
Litza 6338 -	14.03
Fandango 12,908	14.03
Romp Ogden 3d 5458	14.01
Bellita 4453	14.00
Elite 4299 - -	14.00
La Rosa 10,078	14.00

3d sire of

Fadette of Verna 3d 11,122	22.08½
Fairy of Verna 2d 10,973	20.03¾
Hilda A. 2d 11,120 -	20.00
Gardinier's Ripple 11,693 -:	19.12½
Tenella 2d 19,521 -	18.12
Attractive Maid 16,925	16.13
Percie 14,937	16.12
Gazella 3d 9355	16.03
Euphonia 6783 -	16.00½
Thorndale Belle 3d 10,459 -	15.15
Rupertina 10,409	15.12½
Mitten 13,368 -	15.11
Reception 3d 11,025 -	15.08½
Alfritha 13,673 -	15.03
Romping Lass 11,021	15.00
Belle Thorne 13,369	14.11
Euphorbia 11,229 - -	14.09½
Genevieve Sinclair 11,167	14.09
Jaquenetta 10,958 -	14.06
Lady Clarendon 3d 17,578	14.05½
Lottie Rex 18,757 -	14.04
Signetilia 16,333 - -	14.03½
Variella of Linwood 10,954	14.01
Sadie's Choice 7979 -	14.00

4th sire of

Atlanta's Beauty 12,949	21.03 -
Signetelia 16,333 -	14.03½

MARK TAPLEY 270.

Son of Sam Weller 271 and Meg 673.

2d sire of

*Katie Kohlman 7270 -	23.10
Lily of Burr Oaks 11,001	15.13

3d sire of

*Effie 885 A. H. B. - -	22.12
Jersey Queen of Barnet 4201 A. H. B.	19.12
Snowdrop F. W. 16,948	14.08
*Eva 883 A. H. B.	14.00

MARMION 359.

Sire on I. of Jersey, dam Sylph 615.

1st sire of

Pet Anna 1608	14.00

2d sire of

Welma 5942 -	17.08
Mamie Coburn 3798 -	17.08
*Belle of Inda 3867	15.01½
Rose of Hillside 3866 -	14.03½
Elinor Wells 12,068	14.00

3d sire of

*Queen of Maple Dale Farm 7036	14.14
Marpetra 10,284 - -	14.06

MARPETRO 3352.

Son of Marius 760 and Pet Anna 1608.

1st sire of

Marpetra 10,284 -	14.06

MARQUIS 1401.

Sire on I. of Jersey, dam Magnolia 2543.

1st sire of

Meines 3559	14.00

2d sire of

Meines 3d 7741 - ·	20.01

MARS 95.

Son of Jupiter 93 and Edna 807.

1st sire of

*Edith 4th 817	14.00

2d sire of

Hebe 3d 3615 -	15.00
Bessie Bradford 7269	14.02
Rioter 2d's Venus 3658	14.00

3d sire of

Myra 2d 6289 - -	16.00
Bessie Bradford 2d 7271	15.02

MARWELL 2118.

Son of Marius 760 and Annie Wells 1947.

1st sire ot

*Queen of Maple Dale Farm 7036	14.14

MATCHEM 747.

Son of Mercury 432 and Azile 1256.

2d sire of

Lily of Maple Grove 5079	16.03

3d sire of

Lida Mullin 9198	16.08
Lizzie D. 10,408	14.00

MATCHLESS 906.
3d sire of

Zittella 2d 11,922	17.08½
Malope 2d 11,923	15.10

MAY BOY 705.
Son of Bismarck 292 and Crocus 1787.
2d sire of

Hazen's Nora 4791	20.04
Silveretta 6852 -	16.09
Princess Shiela 7297	16.04½
Alhena 15,995	16.03
Tobira 8400	15.13
Orphean 4636 -	15.07
Champion Chloe 12,255	15.05½
Dairy C. 12,227	15.01
Coronilla 8367	14.09½
Maggie C. 12,216	14.06
Maggie May 2d 12,926	14.06
Gilt Edge C. 12,224 -	14.03½
Minnie Lee 2d 12941	14.03
Webster's Pet 4103	14.02
Therese M. 9364 -	14.02
Jessie Leavenworth 8248	14.00

3d sire of

Belmeda 6229 - -	18.12
Lady Gray of Hill Top 2d 14,641	14.12
Bell Rex 11,700	14.10
Kate Daisy 8204 - -	14.04
Lady Gray of Hill Top 3d 14,642	14.02
Hurrah Pansy 12,153	14.01½
Baby Buttercup 10,888 -	14.00

4th sire of

Hillside Gem 16,640 -	20.00
La Pera 2d 13,404	14.08

MAY DUKE 90.
4th sire of

Vieva 3d 7642	16.05

MAXSE 400.
2d sire of

Sultana 2d 11,798	15.04

3d sire of

Goldthread 4945 -	16.09

McCLELLAN 25.
Son of Capt. Darling 535 and Angelina Baker 13.
2d sire of

Lady Mel 2d 1795 -	21.00
Champion Chloe 12,255	15.05½
Lady Brown 433 -	14.00
*Allen's Fawnette 3722 - -	14.00
*Brightness 2211	14.00

3d sire of

Value 2d 6844	25.02½½
Peggy Leah 3097	18.12
Kitty Potter 9893	18.05
Maggie 3d 3221	17.08
Dimple 3248 -	16.11
Cascadilla 3103	15.12
*Filbert 3630 -	15.12
*Kitty Lake 8250	15.08½
*Brightness 3d 14,824	15.05
Romp Ogden 2d 4764	15.05
Lady Brown 4th 6911	14.12
*Churchill's Betsey 4105	14.00

4th sire of

Lady Mel 2d 1795	21.00
May Blossom 5657	18.11
Maggie 3d 3221	17.08
Jersey Cream 3151	17.00
Arawana Queen 5368	16.09
Silveretta 6852 -	16.09
Princess Shiela 7297 -	16.04½
Genevieve Sinclair 11,167	16.02
Tobira 8400 - -	15.13
Champion Chloe 12,255 -	15.05½
Forget-Me-Not-O. 10,564	15.04
Dairy C. 12,227	15.01
Olie 4133 -	15.00
Mary Clover 9998	14.15
Bloomfield Lady 6912	14.12
Coronilla 8367	14.09½
Deborana 4718 -	14.08
Maggie May 2d 12,926	14.06
Maggie C. 12,216	14.06
Gilt Edge C. 12,223	14.03½
Minnie Lee 2d 12,941	14.03
Webster's Pet 4103	14.02
Therese M. 8364	14.02
Creamer 2467	14.01
Romp Ogden 3d 5458	14.01
Jessie Leavenworth 8248	14.00
Baby Buttercup 10,888 -	14.00

McCLELLAN 323.
3d sire of

Ebon Edith 10653	19.01
Perces Lee 5538 -	16.10
Snowdrop F. W. 16,948	14.08

4th sire of

Niva 7523	15.08

McCLELLAN 3D 27.
Son of McClellan 25 and Jenny 16.
3d sire of

Lady Mel 2d 1795	21.00
Jersey Cream 3151	17.00
Olie 4133 -	15.00
Creamer 2467	14.01

4th sire of

Peggy Leah 3097	18.12
May Blossom 5657	18.11
Alfleda 6744 -	16.04
Genevieve Sinclair 11,167	16.02
Duchess of Argyle 8758	14.13
Cowle's Nonesuch 6199 -	14.12
Jersey Cream 2d 8519	14.12

McCLELLAN 4TH 85.
Son of McClellan 25 and Lily 1.
1st sire of

Lady Brown 433	14.00

2d sire of

Lady Mel 2d 1795	21.00
Kitty Potter 9893	18.05
*Filbert 3630 -	15.12
Lady Brown 4th 6911 -	14.12

3d sire of

Arawana Queen 5368	16.09
Genevieve Sinclair 11,167	16.02
Bloomfield Lady 6912	14.12

4th sire of

Alhena 15,995	16.03
Louvie 3d 6159 - -	14.13

McCLELLAN 6TH 181.

Son of McClellan 25 and Pansy 6th 38.

2d sire of

Maggie 3d 3221	17.08

3d sire of

Maggie 3d 3221 -	17.08

4th sire of

Cowle's Nonesuch 6199	14.12
*Myra Overall 10,317 -	14.00

MEDWAY 717.

Son of Mr. Micawber 556 and Nellie 289.

1st sire of

Mirtha 3437	17.13½
Medrena 3939	17.12

3d sire of

Medrie Le Brocq 8888	- 14.07

MERCURY 432.

Son of Jupiter 93 and Alphea 171.

1st sire of

*Locusta 5143	21.07
Phædra 2561 -	19.13
Nymphæa 5141 -	18.07½
Lerna 3634	15.12
Zalma 8778	15.05
Purest 13,730 -	15.04
Clytemnestra 2455	15.03½
Reality 16,537	15.03½
Iola 4627 -	15.02½
Marvel 13,734	15.01
Nimble 22,335 - -	14.10
Smoky 13,733	14.09
Renown 13,729 -	14.06
Ideal Alphea 18,755	14.06
Richness 16,536 -	14.06
Alphea Star 16,532 -	14.04½
Alphetta 16,531	14.02½
Vestina 2458	14.02
Lernella 22,322 -	14.01½
Ballet Girl 18,750 -	14.01
Alphea Jewel 22,331	14.00

2d sire of

*Katie Kohlman 7270	23.10
Nymphæa 5141	18.07½
Colie 8309 -	18.04
Zitella 2d 11,922	17.08½
Lass Edith 6290 -	17.00
Clytemnestra 2d 5868	16.00
*Myra 2d 6289	16.00
Bessie S. 5002 -	16.00
Fillpail 16,530	15.11
Malope 2d 11,923	15.10
Idalene 11,841	15.08½
Crust 4775 -	15.07
Purest 13,730	15.04
Faustine 10,354	14.14½
Ideal 11,842 -	14.12½
Hartwick Belle 7722 -	14.08
Richness 16,536 - -	14.06
Alphetta 16,531 -	14.02½
Bessie Bradford 7269	14.02
Lernella 22,322	14.01½
Honey Drop 10,033	14.00½
St. Nick's Flora 16,195 -	14.00

3d sire of

Little Torment 15,581	23.02½
Bomba 10,330 -	21.11½
Zittella 2d 11,922 .	17.08½
Lily of Maple Grove 5079	16.08
Corn 10,504 - -	16.02
Malope 2d 11,923 -	15.10
Niva 7523 - -	15.08
Purest 13,730	15.04
Nazli 10,327 . -	15.03½
Bessie Bradford 2d 7271	15.02
Forsaken 7520	15.01
Marvel 13,734	15.01
Smoky 13,733	14.09
Renown 13,729 -	14.06
Ideal Alphea 18,755	14.06
Alphea Star 16532	14.04½
Robinette 7114 -	14.01
*Lady Golddust 2d 19,861	14.01
Alphea Jewel 22,333	14.00

4th sire of

Lida Mullin 9198	16.08
*Pedro Alphea 13,889	15.05
Reality 16,587 -	15.03½
Alphea Star 16,532 -	14.04½
Lizzie D. 10,408	14.00

MERRY ANDREW 719.

Son of Monarch of Roxbury 499, and Mirth 92.

1st sire of

Merry Burlington 7600	15.04

MERRY BOY 61 J. H. B.

Son of Stockwell 2d 24 J. H. B. and Eva F. 628 J. H. B.

2d sire of

Prize Clementine 10,322	15.12
Forget-me-not 5809	15.08
Deerfoot Girl 15,329	15.08
Energy 22,016	14.05

3d sire of

*Punchinello 11,875	17.11½
Lydia of Libby 11,698	15.03
Forsaken 7520	15.01
Sweetrock 18,256	14.11½
Jazel's Maid 11,011	14.06

4th sire of

St. Jeannaise 15,789	16.04
Sweet Sixteen 10,682	14.15

METACOMET 215.

Son of Cliffs 290 and Bonnie 369.

3d sire of

Milkmaid of Burr Oaks 9035	14.05

METEOR 453.

Son of Cliff 176 and Gilt 1176.

1st sire of

Estrella 2831 -	14.12

3d sire of

Florry Keep 6556 .	14.14

MIANTONOMOH 730.

Son of Duke of Wellington 608 and Jura 224.

3d sire of

Katie Bashford 15,982	17.00

4th sire of

Hillside Gem 16,640	20.00

MICAWBER 4796.

Son of Mr. Micawber 556 and Rosa of Lakeside 2d 10,241.

1st sire of

Clara of Lakeside 10,827	-	15.00
Dolly of Lakeside 10,824		14.08

2d sire of

Clara of Lakeside 10,827	15.00

MILO 590.

Son of Lawrence 61 and Motto 80.

1st sire of

Allie Minka 2982	14.06½
Cigarette 2849	14.04
Muezzin 3670	14.00

2d sire of

Bonnie Yost 7943	18.02
Ada Minka 15,562	14.00

MIRABEAU 3800.

Son of Browny 158 J. H. B. and Blossom of the Grange 6958.

1st sire of

Variella of Linwood 10,954	14.01

MODESTE'S MAXSE 1093.

Son of Maxse 400 and Modeste 2627.

4th sire of

Favorite's Rajah Rex 16,153	15.00

MOGUL 532.

2d sire of

Mary M. Allison 6308	-	20.14
Rosebud of Allerton 6352		19.12
Leonice 2d 8842	-	16.08
Lady Alice of Hillcrest 7450		16.03
Glendelia 10,524	-	15.12
Merry Burlington 7600	-	15.04
Violet of Glencairn 10,221		14.04
Queen of Prospect 11,997		14.02

3d sire of

Rosebud of Allerton 6352	-	19.12
Lady Josephine 11,560 (8 days)		19.02

4th sire of

Thorndale Belle 3d 10,459	15.15
Mitten 13,368	15.11
Alfritha 13,673	15.03
Belle Thorne 13,369	14.11

MOGUL 568.

2d sire of

Calendine 9415	17.09
Mary of Gilderoy 11,219	14.04

3d sire of

Little Torment 15,581	23.02½

MONADNOCK 258.

Son of Bashan 146 and Lady Webster 638.

2d sire of

Lady Penn 5314	-	16.00

3d sire of

Goldthread 4945	16.09

MONARCH 82.

2d sire of

Blanche 594	16.00

3d sire of

Ianthe 4562	19.11
Cyrene 4th 480	17.01
Belle of Middlefield 1516	16.03
Arietta 5115	15.00
Alice of Salem 5053	14.08

4th sire of

Chroma 4572	20.06
Pyrola 4566	18.06
Cerita of Meadowbrook 5056	17.08
Silver Rose 4753	16.14
Pattie Mc. 3d 4754	16.08
Zithey 9184	16.07
*Nipheta 9180	16.00
Princess Bellworth 6801	15.10½

MONARCH 165.

1st sire of

Pansy 602	14.00

2d sire of

Helve 4565	14.00

3d sire of

Joan d'Arc 2163	16.13½

4th sire of

Nannie Fitch 9143	14.04

MONARCH OF ROXBURY 499.

Son of John Le Bas 398 and Nellie 289.

1st sire of

Topsey Roxbury 7796	15.00
Audrey 1447 -	14.00
Merry Burlington 7600	15.04
Clover Top 9910 -	14.00

3d sire of

Walkyrie 5708	14.01

MONITOR 878.

Son of Rob Roy 17 and Emma 801.

1st sire of

Belle Grinnelle 4073	18.08
White Clover Leaf 4512	17.15

2d sire of

Belle Grinnelle 3d 16,503	14.02

3d sire of

Grinnelle Lass 11,859	16.10
Lady Cecilia 24,821	16.01
Prince's Bloom 9729	14.03

MONMOUTH 210.

Son of Earl 81 and Lilac 340.

1st sire of

Cyrene 4th 480	17.01

2d sire of

Cerita of Meadowbrook 5056	17.08
Pattie Mc 3d 4754	16.08

3d sire of

Beeswax 9807	17.05
Busy Bee 6336 -	16.04
Blossie Reynolds 6082	16.03½
Charmer 4771 -	14.12
Home Matron 6707	14.00

4th sire of

Rosebud of Allerton 6352 . -	19.12
Lady Josephine 11,560 (8 days) -	19.02
Maudine of Elmwood 8718 -	16.15
Alhena 15,995	16.03
Nelida 2d 8227 -	15.02½
Royal Sister 12,457	14.11

MONSIEUR 1723.

Son of Vespucius 758 and Madame 1812.

1st sire of

Azelda 2d 7022	15.02

MOPSUS 1165.

Son of Dolphin 2d 468 and Julie Richards 1674.

1st sire of

Miss Willie Jones 6918	16.04
Faustine 10,354	14.14½

2d sire of

Marvel 13,734 -	15.01

MORLEY 644.

Son of Colonel 643 and Flirt 1632.

4th sire of

Bet Arlington 8970	18.11

MORSE 847.

Son of Vanguard 845 and Cowslip of Tonawanda 2116.

2d sire of

Florinanna 9862	17.05
Vieva 3d 7642	16.05

MOSCOW 2303.

Son of Vermont 893 and Magna 5th 3541.

1st sire of

Lady Adams 2d 6529	15.03

MOTLEY 515.

Son of Jack Horner 514 and Meg Merrilies 1372.

2d sire of

Mink 2d 3890 -	19.11
Oktibbeha Duchess 4422	17.04
Arawana Poppy 6053	15.02
Mink 3d 4868	14.09
Adora 18,569	14.03

3d sire of

Mhoon Lady 6560	17.03
Ochra 2d 11,516 -	16.06½
Julia Evelyn 6007	15.15½
Valerie 6044 -	15.13
Dairy C. 12,227	15.01

4th sire of

Marpetra 10,284 -	14.06
Therese M. 8364	14.02

MR. GUPPY 993.

Son of Brutus Woodford 703 and Lady 1775.

1st sire of

Hazalena's Butterfly 10,123	14.00

MR. MICAWBER 556.

1st sire of

Countess Micawber 1759 .	16.08

2d sire of

Mirtha 3437	17.13½
Medrena 3939 -	17.12
Clara of Lakeside 10,827	15.00
Dolly of Lakeside 10,824 -	14.08
Woodland Lass 3444	14.00

3d sire of

Mink 2d 3890 -	19.11
Clara of Lakeside 10,827	15.00
Mink 3d 4868 -	14.09
Village Maid 7069	14.00

4th sire of

Mhoon Lady 6560	17.03
Julia Evelyn 6007	15.15½
Medrie Le Brocq 8888	14.07

MR. TOODLES 377.

3d sire of

Jenny Dodo H. 14,448	21:08
Nibbette 11,625 .	14.07

MUCH ADO 2405.

Son of Dash of Glastonbury 1959 and Dandelion 2521.

1st sire of

Roll of Honor 13,610 -	14.12

MUSIC 118 J. H. B.

1st sire of

Bella Delaine 10,356 -	14.02

NAPOLEON 225.

Son of Major 75 and Brenda 789.

2d sire of

Judy 691 -	19.00
Mary Clover 9998	14.15

3d sire of

Tilda 3720	15.00

4th sire of

Peggy Leah 3097	18.12
Summerline 8001 -	18.06
Duchess of Argyle 3758	14.13
Topsey K. 22,769 -	14.00

NAPOLEON 291.

1st sire of

Gipsey 5th 2252	17.02

2d sire of

Peggy of Staatsburg 2342	14.01½
Daisy of Clermont 3492	14.00

3d sire of

Empress 6th 3203	-	17.09¾
Goddess of Staatsburg 5252		14.08
Celia Belle 5865		14.03

4th sire of

Pride of Bovina 8050		16.09
Dorothy of Bovina 9373		15.04
Lady Adams 2d 6529		15.03
Miss Baden Baden 14,760		14.14½
Queen of De Soto 12,318		14.13
*Myra Overall 10,317		14.00

NAPOLEON 2d 527.
Son of Major 378 and Europa 558.
1st sire of

| Topsey K. 22,769 | | 14 00 |

2d sire of

| Miss Blanche 2515 (10 days) | | 20.09 |

NARRAGANSETT 375.
Son of Challenger 376 and Flora 2d 979.
3d sire of

| Polly Clover 7052 | | 16.15 |

4th sire of

| Olie's Lady Teazle 12,307 | | 16.05 |

NARRAGANSETT 536.
Son of Comet 130 and Diana 261.
2d sire of

| Nellie Maitland 4450 | | 16.00 |

NED 20.
Son of Shaker 21 and Fanny 72.
2d sire of

| Myrtle 2d 211 | | 15.12 |

3d sire of

| Myrtle of Ridgewood 7858 | | 14.01 |

4th sire of

Belmeda 6229		18.12
Lida Mullin 9198	-	16.08
Rosabel Hudson 5704		15.12
Canto 7194	-	15.12
Little Sister 11,666		14.12
Lady Ives 3d 6740		14.08
Aroma 8518	-	14.07
Fandango 12,908		14.03
Miami Prize 8100		14.00
Lizzie D. 10,408		14.00

NED 523.
1st sire of

| Victoria 3175 | | 16.01 |

3d sire of

| Gold Lace 10,726 | | 14.13 |

4th sire of

| Gold Mark 10,727 | | 14.14 |

NED BOOTH 1508.
Son of Arab 245 and Hortense 3739.
3d sire of

| Alfieda 6744 | - | 16.04 |

4th sire of

| Alfritha 13,673 | | 15.03 |

NED IVES 3631.
Son of Beeswax 1931 and Lady Ives 1708.
2d sire of

| Percie 14,937 | | 16.12 |

NELSON 65 J. H. B.
2d sire of

| Miss Browny 7288 | · | 16.13 |

NELUSKO 479.
Son of Rajah 340 and Nelly 55.
1st sire of

Belle of Milford 7445		14.07
Maggie May 3255		14.02½
Gilt Edge 2d 4420		14.00

2d sire of

Lucky Belle 2d 6037		16.14
Julia Evelyn 6007		15.15½
Valerie 6044	-	15.13
Duchess Caroline 3d 6039		15.08
Bettie Dixon 4527		15.00
Florry Keep 6556		14.14
Coronilla 8367	-	14.09½
Pride of the Hill 4877		14.08
Maggie C. 12,216		11.06
Maggie May 2d 12,926		14.06
Starkville Beauty 4897		14.00

3d sire of

Atlanta's Beauty 12,949		21.03
Armon 10,862	-	16.13½
Mountain Lass 12,921	·	14.09
Gilt Edge C. 12,223		14.03½
Minnie Lee 2d 12,941		14.03
Therese M. 8364		14.02
Vivalia 12,760		14.00

4th sire of

| Marpetra 11,284 | | 14.06 |

NEPTUNE 14 J. B. H.
Son of Brown Prince 85 J. H. B. and Fan F. 207 J. H. B.
1st sire of

| Coomassie 11,874 | | 16.11 |

3d sire of

Princess 2d 8046		46.12½
Ona 7840	-	20.13
*Punchinello 11,875		17.11½
Young Garenne 13,641		17.08
Lady Velveteen 15,771		17.02
Les Cateaux 2d 15,538		16.01
Lady Kingscote 26,085		15.10
Lady Vertumnus 13,217		14.10
Blonde 2d 9268	-	14.04
La Rouge 12,405		14.02
Daisy Queen 9619		14.00
Gazelle 15,961	-	14.00
Lady Young 16,688		14.00

4th sire of

Oxford Kate 13,646		39.12
Little Torment 15,581		23.02½
Island Star 11,876		21.03
Daisy Brown 12,203	-	17.06
Princess of Ashantee 13,467		16.05

St. Jeannaise 15,789   -   16.04
Mabel of Trinity 13,694   16.03
Desire 9654 - -   16.03
Odelle Sales 15,564   16.03
Rose of Oxford 13,469   -   15.14½
Queen of Ashantee 14,554   15.02
Romping Lass 11,021   15.00
Auntybel 12,582 -   14.09
Ada Minka 15,562   14.00

NEPTUNE 230 J. H. B.

Son of Noble 2d 256 J. H. B. and Dora 155 J. H. B.
1st sire of

Dora Neptune 20,318 '-   20.00½

NEPTUNE 842.

Sire on I. of Jersey, dam Lady Mary 2104.
2d sire of

Rosa of Glenmore 3179 -   17.10
Embla 4799   -   17.08

NEPTUNE OF THE GRANGE 1549.

Sire on I. of Jersey, dam Jersey of the Grange 3807.
4th sire of

*Pedro Alphea 13,889   -   15.05

NERO 13.

4th sire of
Lottie Rex 18,757   14.04

NERO CHIEF 2951.

Son of Pierrot 5th 1665 and Mite 2751.
2d sire of

Olie's Lady Teazle 12,307   16.05

NERO CHIEF 4217.

Son of Nero Chief 2951 and Dolly Newell 4254.
1st sire of

Olie's Lady Teazle 12,307   16.05

NESTOR 773.

Son of Albert 44 and Lady Mel 429.
2d sire of

Louvie 3d 6159   - -   14.13

NESTOR 1834.

Son of Prince John 22 and D. Buck's Clover 20.
1st sire of

*Belle of Bloomfield 4331   14.00
2d sire of
Rosa Miller 4333   - - -   17.07
3d sire of
Rosa Miller 4333 -   17.07
Bloomfield Lady 6912   14.12

' NEW WORLD 289.

Sire on I. of Jersey, dam Ella 767.
3d sire of

Beauty Bismarck 4967   14.01

NIMROD 28.

Son of Beckwith's Bull 29 and Young Duchess 41.
3d sire of

Medrena 3939   17.12
Lara 4309   -   17.08
Rene Ogden 1568   -   15.00
Kalmia 4561   15.00
4th sire of
Jeannette Montgomery 5177   20.00
Safrano 4568   17.08
Kaoli 18,980   17.08
Renalba 4117   -   17.04½
Mhoon Lady 6560   17.03
Auria 4561 -   16.18
Silenta 17,685   15.10
Calypris 5943   -   15.09½
Florry Keep 6558 -   14.14
Mountain Lass 12,921   14.09
Bintana 9837   14.03½
Erith 4564   14.00
Jule 8640   14.00
Silene 4307   14.00

NIMROD 2D 246.

Son of Nimrod 28 and Kitty Clyde 30.
2d sire of

Lara 4306   17.08
3d sire of
Jeannette Montgomery 5177   -   20.00
Lutea 4563   -   16.03
Silenta 17,685   15.10
Silene 4307   -   14.00
Jule 8640   14.00
4th sire of
Jeannie Platt 6005   -   14.04

NIOBE DUKE 2364.

Son of Jeweler 1385 and Niobe 6th 8516.
2d sire of

Thorndale Belle 3d 10,451   15.15
Mitten 13,368   15.11
Alfritha 13,678   -   15.08
Belle Thorne 18,369   14.11

NIOBE GRAND DUKE 4510.

Son of Niobe Duke 2364 and Witch Hazel 3d 4875.
1st sire of

Thorndale Belle 3d 10,451   15.15
Mitten 13,368   -   15.11
Alfritha 13,673   -   15.03
Belle Thorne 18,369   14.11

NOBLE 104 J. H. B.

Son of Sultan 58 J. H. B. and Handsome 265 J. H. B.
1st sire of

Regina 2d 2475   - -   14.08

2d sire of

Chrome Skin 7881 -	20.10
Queen of Delaware 17,029	18.13
Garenne 1575 J. H. B. -	17.08
Cream of Sidney 17,028 -	17.02½
Walkyrie 5708 -	14.01

3d sire of

*Sultane 2d 11,373 -	23.08
Daisy of St. Peters 18,175 -	20.05½
Oakland's Cora 18,853	19.09½
Reception 8557 -	19.08
Queen of Delaware 17,029	18.13
Merry Duchess 13,963 -	18.09½
Panatella 4778 -	18.03
Calendine 9415 - -	17.09
Faith of Oaklands 19,696	17.04
Fear Not 6059 -	17.03
Brunette Le Gros 9755	15.15
Lucilla Kent 8892 -	15.10
Fan of Grouville 7458	15.00
Jenny Le Brocq 9757	14.14
Cocotte 11,958 -	14.12
Satin Bird 16,380 -	14.10
Daisy of Chenango 18,582	14.07

4th sire of

Princess 2d 8046	46.12½
Ona 7840 - - -	20.13
Daisy of St. Peters 18,175 -	20.05½
Dora Neptune 20,318	20.00½
Oakland's Cora 18,853	19.09½
Floribundus 2d 14,949	18.08
Lactine 10,680 -	17.01½
St. Jeannaise 15,789	16.04
Fear Not 2d 6061 - -	16.02
Dairy Pride 4th 521 J. H. B.	16.00
Brunette Le Gros 9755	15.15
Prize Clementine 10,322	15.12
Lucilla Kent 8892 -	15.10
Deerfoot Girl 15,329	15.08
Forget-me-not 5809 - -	15.08
Beauty 17,414 -	15.00
Jenny Le Brocq 9757	15.14
Cocotte 11,958 -	14.12
Cosetta 15,991	14.11
Energy 22,016	14.05
Blonde 2d 9268	14.04
Ballet Girl 18,750	14.01
Daisy Queen 9619	14.00

NOBLE 901 (195 J. H. B.).

Son of Noble 104 J. H. B. and Fanny of Babylon 2345.

1st sire of

Queen of Delaware 17,029	18.13
Cream of Sidney 17,028	17.02½
Desire 9654 -	16.03

2d sire of

Queen of Delaware 17,029	18.13
Cream of Sidney 17,028	17.02½

NOBLE 2D (256 J. H. B.).

Son of Welcome 166 J. H. B. and Lily 23 J. H. B.

2d sire of

Dora Neptune 20,318	20.00½

NONPAREIL 37 J. H. B.

Son of Orange Peel 129 J. H. B. and Les Cateaux F. 487 J. H. B.

2d sire of

Young Garenne 13,641 -	17.08
Gazelle 15,961	14.00

3d sire of

Island Star 11,876 - -	21.03
Princess of Ashantee 13.467	16.12
Mabel of Trinity 13,694	16.03
Queen of Ashantee 14,554	15.02
Auntybel 12,582	14.09

4th sire of

*Pendule 2d 16,709 -	16.03
Queen of Ashantee 2d 16,657	14.03½

NONQUIT 1391.

Son of Motley 515 and Primula 1288.

2d sire of

Ochra 2d 11,516 - -	16.06½

NORAJAH 812.

Son of Rajah 340 and Nora 434.

1st sire of

Hazen's Nora 4791 -	20.04
Arawana Buttercup 6052	15.05
Arawana Poppy 6053	15.02

2d sire of

Arawana Queen 5368	16.09

NORMAN 73.

4th sire of

Lady Love 2d 2212	16.08
Helen 3556 -	15.09
My Queen 12,614 -	15.08
Del of Willow Farm 22,464	14.08
Meines 3559	14.00

NORMANDY 1046.

Son of Mack 722 and Mischief 954.

2d sire of

Kate Gordon 8387	15.15
*Myra Overall 10,317	14.00

NORWOOD 1077.

Son of Rosa 633 and Norina 1929.

1st sire of

Goldthread 4945	16.09

NUTSHELL 729.

Son of Neluska 479 and Fanny Ogden 1564.

1st sire of

Pride of the Hill 4877	14.08

2d sire of

Vivalia 12,760 :	14.00

NUTSHELL, JR., 1500.

Son of Nutshell 729 and Guinevere 1484.

1st sire of

Vivalia 12,760	14.00

NYE 667.
Son of Monmouth 210 and Niobe 99.
2d sire of

Blossie Reynolds 6082	16.03½
Home Matron 6707	14.00

3d sire of

Royal Sister 14,457 - -	14.11

4th sire of

Atlanta's Beauty 12,959	21.03

OAK BLUFFS 908.
Son of Monarch of Roxbury 499 and Jenny May 473.
1st sire of

Clover Top 9910	14.00

OAKLAND 33.
Son of Com. Nutt 36 and Cowslip 43.
2d sire of

*Optima 6715 -	23.11
Romp Ogden 2d 4764	' 15.05

3d sire of

Romp Ogden 3d 5458	14.01

4th sire of

Romping Lass 11,021	15.00
Cowle's Nonesuch 6199	14.12

OKUBO 1876.
Son of Lawrence 61 and Maize 1588.
1st sire of

Irene of Short Hills 5737	14.06½

"OLD NOBLE" (I. OF J.).
1st sire of

Regina 32 J. H. B.	20.04
Soucique 68 J. H. B.	16.00

2d sire of

Regina 4th 12,732	17.13½
Regina 2d 2475	14.08

3d sire of

Lisette 483 J. H. B.	16.04
Walkyrie 5708 -	14.01
Fille de l'Air 2474	14.00

4th sire of

Nelly 6456	21.00
Reception 8557 -	19.08
Garenne 1575 J. H. B.	17.08
Fear Not 6059 - -	17.03
Dairy Pride 4th 521 J. H. B. -	16.00
Dairy Pride 6th 21,681	16.00
Atricia 6029	15.03
Regina 2d 2475	14.08
Esperanza	14.03

OMAHA 482.
Sire on I. of Jersey, dam Omoo 1247.
1st sire of

Metah's Queen 4886	17.09
Bryant 4193 -	14.08

ONECO 918.
Son of Uncas 628 and Young Fancy 97.
1st sire of

Rosabel Hudson 5704	15.12

ONTARIO 865.
Son of Black Imperial 255 and Helen 2180.
1st sire of

Maple Leaf 4768	14.12

2d sire of

Oak Leaf 4769	17.10

3d sire of

Bet Arlington 8970	18.11
Euphorbia 11,229 -	14.09½

4th sire of

*Lenoreisa 16,233 -	21.00
Gardinier's Ripple 11,693	19.12½
Countess Queen 13,519	18.03
Countess Lowndes 26,874	17.08
Litty 8017 -	14.00

OPTIMUS 1607.
Son of Sir Charles 131 and Carrie 3894.
1st sire of

Warren's Duchess 4622 -	16.01

2d sire of

Dot of Bear Lake 6170 ·	19.04
Mary of Bear Lake 6171	14.07

3d sire of

Countess Lowndes 26,874	17.08

ORANGE BUD 2978.
Son of Orange Skin 1216 and Brown Gipsey 2d 5095.
2d sire of

Corn 10,504 ·	16.02

ORANGE PEEL 502 (129 J. H. B.)
Son of Clement 115 (61 J. H. B.) and Cowslip 330 J. H. B.
1st sire of

Lustre 2061	15.08½

2d sire of

Daisy Pride 4th 521 J. H. B.	16.00
Daisy Pride 6th 21,680	16.00
Rose of Oxford 13,569	15.14½
Atricia 6029 -	15.03
Rose 2587 J. H. B.	14.10
Gilda 2779 · -	14.06

3d sire of

Rosa of Belle Vue 6954	18.07½
Young Garenne 13,641	17.08
Mary Jane of Belle Vue 6956	17.07
Viva Le Brocq 13,702	17.07
Gold Trinket 9518 -	16.02
Lilly of Burr Oaks 11,001	15.13
Belle Dame 2d 22,043	15.03
Grace Feich 8291 - -	15.00
Caroline 12,019 - -	14.08
Milkmaid of Burr Oaks 9035	14.05
Gazelle 15,961	14.00 .
Lizzie C. 7713 -	14.00
Carlo's Fanny 14,951	14.00 .

4th sire of

Jenny Dodo H. 14,448	21.08
Island Star 16,876 - -	21.03
Lady Josephine 11,560 (8 days)	19.02
Mabel of Trinity 13,694 -	16.03

Countess Gasela 9571	- 15.11
Nelida II. 8117 -	15.02½
Queen of Ashantee 14,554	15.02
Naomi's Pride 16,745	15.02
Auntybel 12,582 -	14.09
Bella Delaine 10,356	14.02

ORANGE PEEL 864.
1st sire of

Valma Hoffman 4500	21. 09

2d sire of

Leonice 2d 8842 -	16.08

3d sire of

Nelida 2d 8227	15.02½

ORANGE PEEL 2D 36 J. H. B.
Son of Orange Peel 129 J. H. B.
2d sire of

Rosa of Belle Vue 6954 -	18.07½
Mary Jane of Belle Vue 6956	17.07
Gold Trinket 9518	16.02
Caroline 12,019	14.08

3d sire of

Countess Gasela 9571	15.11
Naomi's Pride 16,745	15.02
Bella Delaine 10,356	14.02

4th sire of

Corn 10,504	16.02

ORANGE SKIN 19 J. H. B.
Son of Noble 104 J. H. B. and Longueville Queen 272 J. H. B.
2d sire of

*Sultane 2d 11,373	23.08
Merry Duchess 13,693	18.09½
Panatilla 4778	18.03

3d sire of

Cosetta 15,991	14.11

ORANGE SKIN 1216 (227 J. H. B.).
Son of Orange Peel 2d 36 J. H. B. and Gloria 3144.
1st sire of

Gold Trinket 9518	16.02

2d sire of

Naomi's Pride 16,745	15.02
Bella Delaine 10,356	14.02

3d sire of

Corn 10,504	16.02

ORAWAMPUM 2833.
Son of Tommy Grey 10,099 and Carrie 7th 2016.
1st sire of

Corn 10,504	16.02

ORI 4286.
Son of Oxoli 1922 and Chenie 4570.
1st sire of

Flamant 11,270	14.02

2d sire of

Trudie 17,770	14.10

ORION 355.
Son of Junius 204 and Kitty Clyde 30.
3d sire of

Dom Pedro's Julian 8638	16.00

ORLEANS 533.
Son of Warwick 264 and Daisy 656.
1st sire of

Stanstead Belle 4709	14.11½

ORLOFF 3143.
Son of Lord Lisgar 1060 and Ophelie 493.
1st sire of

Carrie Pogis 22,568	15.09

3d sire of

Niobe of St. Lambert 12,969	21.09½
Rose of St. Lambert 30,426	21.03½
Rioter's Beauty 14,894	14.00

ORPHAN 891.
Son of Napoleon 291 and Olive 763.
1st sire of

Peggy of Staatsburg 2342	14.01½

OSCAR 98.
Son of Comus 54 and Fanny 180.
3d sire of

Cornucopia 3414	15.12

OSSIPEE 679.
Son of Bashan 146 and Islip 1764.
2d sire of

Lady Penn 5314	16.00

OTHELLO 1114.
Son of Tally Ho 880 and Theodora 1896.
1st sire of

Dark Cloud 9364	15.03½

OURI 2916.
Son of Champion of America 1567 and Kate Nickleby 3100.
2d sire of

Hillside Gem 16,640	20.00

OXOLI 1922.
Son of St. Helier 45 and Pyrola 4566
1st sire of

Volie 19,465	- 18.01
Kaoli 18,980	17.08
Zithey 9184	16.07
Lesbie 9179	- 16.03
*Nipheta 9180	- 16.00
Renini 9181	14.10½
Bintana 9837 -	- 14.03½
Taglioni 9182 -	14.01

2d sire of

Trudie 17,770 -	14.10
Queen of Chenango 17,771	14.06
Flamant 11,270	- 14.02

3d sire of

Trudie 17,770	14.10

PADDY 899.
1st sire of

Cherry 3d (unreg.)	19.04½

2d sire of

Cream of Sidney 17,028	17.02½

3d sire of

Queen of Delaware 17,029	18.13

PADDY WILSON 3084.
Son of Burnside 2d 2838 and Lilac Hanmer 4382.
1st sire of

Lady Bidwell 10,303	15.12

PADISHA 1623.
Son of Rajah 340 and Grisette 596.
1st sire of

Pet Lee 7993	14.12

PAGAN 1800.
Son of Young Pilgrim 302 and Eye Bright 1331.
2d sire of

Bet Arlington 8970	-	18.11

3d sire of

Countess Queen 13,519	18.03
Litty 8017	14.00

4th sire of

*Lenoreisa 16,233	-	21.00

PAN 110.
Son of John Brown 67 and Pansy 2d 259.
3d sire of

Negress 7651 -	14.00

4th sire of

*Fawnette of Woodstock 3710	17.08
Thorndale Belle 5265	14.08

PANSY 184 J. H. B.
1st sire of

Queen of the North 17,973	14.00

PARTISAN 235.
Son of Pluto 232 and Potowomut 1337.
2d sire of

Mirth's Blanche 19,572	17.13½

PASHA 64 J. H. B.
Son of Rattler 20 J. H. B.
1st sire of

Regina 4th 12,732	-	17.13½
Faultless 12,018	-	17.05½

2d sire of

Fillpail 16,530	- -	15.11

PATERSON 11.
Son of St. Clement 10, dam imported.
1st sire of

Cowslip 5th 849	15.04

2d sire of

Blanche 594	16.00

3d sire of

Value 2d 6844	25.02½½
Ianthe 4562	19.11
Vixen 7591 -	17.06
Oktibbeha Duchess 4422	17.04
Lucky Belle 2d 6037	16.14
*Pulsatilla 7557	16.03
Maggie May 3255	14.02½

4th sire of

Value 2d 6844		25.02½½
*Optima 6715		23.11
Tenella 6712 -		22.01½
Croton Maid 5305	*-	21.11½
Chroma 4572		20.06
Roonan 5133		18.12
Pyrola 4566 -		18.06
Jersey Cream 3151		17.00
Valhalla 5300		17.00
Silver Rose 4753		16.14
Zithey 9184		16.07
Belle of Patterson 5664		16.06
Urbana 5597 -		16.00
*Nipheta 9180	-	16.00
Kate Gordon 8387		15.15
Œnone 8614		15.14
Edwina 6713		15.13
Valerie 6044		15.13
*Filbert 3630 -		15.12
Fanny Taylor 6714		15.12
Aldarine 5301		15.01½
Olie 4133 -		15.00
Maggie C. 12,216		14.06
Maggie May 2d 12.926		14.06
Kate Daisy 8204		14.04
Litza 6338 - - -		14.03
Variella of Linwood 10,954		14.01
Creamer 2467		14.01
Pixie 4115 -		14.00

PATERSON 4TH 2400.
Son of Paterson 11 and Sue 2d 65.
2d sire of

Value 2d 6844	25.02½½

PEDRO 3187.
Son of Domino of Darlington 2459 and Premium of Darlington 5572.
1st sire of

*Pedro Alphea 13,889	15.05

PEQUABOC CHIEF 2662.
Son of Wethersfield 966 and Mary Goodenough 2068.
2d sire of

Olie's Lady Teazle 12,307	-	16.05

3d sire of

*Wabash Girl 14,550	16.00

PERSEUS 622.
Son of Athol 621 and Diana 2d 499.
2d sire of

Urbana 5597	-	16.00

3d sire of

Corn 10,504	-	16.00

4th sire of

| Corn 10,504 | - | - | 16.02 |

PERTINATTI 713.

Son of Pilot, Jr., 141 and Pert 110.

1st sire of

Beauty	-	20.15
Renalba 4117	-	17.04½
Romp Ogden 2d 4764		15.05
Pixie 4115	-	14.00

2d sire of

Roonan 5133	-	18.12
Leoni 11,868	-	18.02
Bonnie Yost 7943	-	18.02
Dudu of Linwood 8336		16.15
Kate Gordon 8387		15.15
Lucetta 6856		14.03

3d sire of

Daisy Brown 12,213		17.06
Dora Doon 12,909		15.00
Fall Leaf 8587		14.08
Adora 18,569		14.03

PERTINAX 1965.

Son of Pertinatti 713 and Roxana 1761.

1st sire of

| Kate Gordon 8387 | 15.15 |

PETERKIN 2451.

Son of King of Fairview 778 and Jersey Cream 3151.

1st sire of

| Katie Bashford 15,982 | 17.00 |

2d sire of

| Hillside Gem 16,640 | 20.00 |

PETER NORMAN 1238.

Son of Yankee 1003 and Princess of Normandy 3190.

2d sire of

| Dorothy of Bovina 9373 | 15.04 |

PHAROS 3552.

2d sire of

| Belle of Scituate 7977 | 16.00 |
| Lass of Scituate 9555 | 15.14 |

3d sire of

| Minnie of Scituate 17,829 | 14.04½ |

PHAROS, JR., 3621.

Son of Pharos 3552 and Blonde 378.

1st sire of

| Lass of Scituate 9555 | 15.14 |

2d sire of

| Minnie of Scituate 17,829 | 14.04½ |

PHENOMENON 1147.

Son of Excelsior of Jersey 949 and Tilly 2408.

2d sire of

| Milkmaid of Burr Oaks 9035 | 14.50 |

PICKWICK 3985.

Son of Jason of Deerfoot 1636, and Villette 1382.

2d sire of

| Percie 14,937 | 16.12 |

PHILIP SHERIDAN 984.

Son of Living Storm 178 and Belle 1225.

1st sire of

| *Churchill's Betsey 4105 | 14.00 |

2d sire of

| Webster's Pet 4103 | 14.02 |

4th sire of

| La Pera 2d 13,404 | 14.08 |
| Kate Daisy 8204 | 14.04 |

PIERCE'S DOM PEDRO 4373.

Son of Tarquin 750 and Sunbeam 392.

1st sire of

| Bohemian Gipsey 17,452 | 14.11 |

PIERROT 636.

1st sire of

*Belle of Saybrook 6875	15.01½
Elsie Brown 4021	14.06½
Geranium 3963	14.00

2d sire of

*Mary Walker 11,303	18.12
Colt's La Biche 6399	17.02½
Polly Clover 7052	-	16.15
Pierrot's Lady Bacon 12,482	16.10
Hattie Douglass 24,960	-	16.05
Lady Cecelia 24,821	16.01
Pierrot's Picture 12,481	16.00
Mineola of Elmarch 8229	-	15.15
Pierrot's Lady Hayes 11,672	15.12
Fanny Taylor 6714	-	15.12
Julia Walker 10,183	15.12
Lady Hayes 10,136	15.12
Canto 7194	15.12
Palestina 4644	-	15.08
Princess of Mansfield 8070	15.02
Bellini's Maid 15,170	15.01½
Little Sister 11,666	-	14.12
New London Gipsey 11,667	14.08
Jennie of the Vale 9553	-	14.06½
Palestine's Last Daughter 12,602	14.06
Lady Fanning 11,169	14.06
Rosy Kate 10,276	-	14.02
Queen Fannie 10,275	14.02
Rarity 2d 7724	-	14.02
Pierrot's Countess 12,480	14.00

3d sire of

Belmeda 6229	-	18.12
Rosy Kate's Rex 13,192	18.08
Kitty Potter 9893	18.05
Percie 14,937	-	-	16.12
Pierrot's Lady Bacon 12,482	16.10
Lida Mullin 9198	-	16.08
Lady Cecelia 24,821	16.01
Pierrot's Picture 12,482	16.00
Lady Hayes 10,136	-	15.12
Pierrot's Lady Hayes 11,672	15.12
Bellini's La Biche 15,091	14.14½
Yellow Locust 10,679	14.10½
Aroma 8518	-	-	14.07
Susie La Biche 3d 15,171	14.06½

Rosy Kate 10,276	14.04
Lizzie D. 10,408 -	14.00
Pierrot's Countess 12,480 -	14.00

4th sire of

Atlanta's Beauty 12,949	21.03
Rosy Kate's Rex 13,192	18.08
Olie's Lady Teazle 12,307	16.05

PIERROT 2D 1669
Son of Pierrot 636 and Dainty 796.
1st sire of

*Mary Walker 11,303	18.12
Julia Walker 10,183	15.12
Palestina 4644 - -	15.08
Princess of Mansfield 8070	15.02
Little Sister 11,666 - -	14.12
New London Gipsey 11,667	14.08
Rosy Kate 10,276	14.04
Queen Fannie 10,275	14.02

2d sire of

Belmeda 6229 - -	18.12
Rosy Kate's Rex 13,192 -	18.08
Pierrot's Lady Bacon 12,482	16.10
Lida Mullin 9198 - -	16.08
Pierrot's Picture 12,481	16.00
Lady Hayes 10,136 - -	15.12
Pierrot's Lady Hayes 11,672	15.12
Aroma 8518 - -	14.07
Rosy Kate 10,276	14.04
Lizzie D. 10,408 -	14.00
Pierrot's Countess 12,480	14.00

3d sire of

Rosy Kate's Rex 13,192	18.08

PIERROT 5TH 1665.
Son of Pierrot 636 and Eugenie 2d 1623.
3d sire of

Olie's Lady Teazle 12,307	-	16.05

PIERROT 7TH 1667.
Son of Pierrot 636 and Pet 811.
1st sire of

Pierrot's Lady Bacon 12,482	16.10
Pierrot's Picture 12,481 -	16.00
Lady Hayes 10,136 -	15.12
Pierrot's Lady Hayes 11,672 -	15.12
Palestine's Last Daughter 12,602 -	14.06
Lady Fanning 11,169 -	14.06
Pierrot's Countess 12,480	14.00

PILOT 3.
Sire on I. of Jersey, dam Juno 120.
2d sire of

Silver Rose 4753 -	16.14
Thisbe 607	15.12

3d sire of

Thisbe 2d 2201	19.01½
Renalba 4117 -	17.04½
Chamomilla 7552	16.10
Lutea 4563 -	16.03
Romp Ogden 2d 4764	15.05
Belle of Vermillion 8798	15.02
Kalmia 4561 -	15.00
*Woodland Margaret 6215	14.10½
Opaline 7590 -	14.10
Adina 1942	14.04
Pixie 4115 - -	14.00

4th sire of

Mary M. Allison 6308	20.14
Roonan 5133 - -	18.12
Mollie Garfield 12,172	18.07
Leoni 11,868 -	18.02
Bonnie Yost 7943	18.02
Kaoli 18,980 -	17.08
Safrano 4568 -	17.08
Cerita of Meadowbrook 5056	17.08
Renalba 4117- -	17.04½
Princess Moster 9700	17.03
Dudu of Linwood 8336	16.15
Auria 4561 -	16.13
Pattie Mc 3d 4754	16.08
Gossip 6165	16.07
Urbana 5597 -	16.00
Kate Gordon 8387	15.15
Cenie Wallace 2d 6557	15.04½
Bellini's Maid 15,170 -	15.01½
Bellini's La Biche 15,091	14.14½
Florry Keep 6556	14.14
Magnibel 7976	14.12
Charmer 4771 -	14.12
Mountain Lass 12,921	14.09
Epigaea 4631 - -	14.07
Susie La Biche 3d 15,171	14.06½
Jaquenetta 10,958	14.06
Bintana 9837 -	14.03½
Gem of Sassafras 8434	14.03½
Lillian Mostar 10,364	14.03
Lucetta 6856 -	14.03
Queen of Prospect 11,997	14.02
Bathsheba 2556 -	14.01
Elmora Mostar 15,955	14.00
Erith 4564	14.00

PILOT 57.
Sire on I. of Jersey, dam Commerce 84.
3d sire of

Zampa 2194	18.00
Haddie 921	16.00

4th sire of

Memento 1913	14.05
Celia Belle 5865	14.03
Muezzin 3670 -	14.00
Naomi Cramer 8628	14.00

PILOT 183 J. H. B.
Son of Khedive 103 J. H. B.
1st sire of

Oxford Kate 13,646 -	-	39.12

PILOT 3549.
Son of Dick Swiveller 276 and Nellie 7825.
2d sire of

Jersey Belle of Scituate 7828	25.03
Dolly of Lakeside 10,824	14.08

3d sire of

Jersey Belle of Scituate 7828	25.03
Belle of Scituate 7977	16.00
Lass of Scituate 9555 -	15.14
Clara of Lakeside 10,827	15.00
Dolly of Lakeside 10,824	14.08
Minnie of Scituate 17,829	14.04½

4th sire of

Belle of Scituate 7977	16.00
Lass of Scituate 9555	15.14
Clara of Lakeside 10,827	15.00
Minnie of Scituate 17,829	14.04½

PILOT BOY 488.
Son of Pilot 3 and Marigold 840.
2d sire of

Chamomilla 7552 -	16.10
Belle of Vermillion 8798	15.02
Adina 1942	14.04

3d sire of

Mollie Garfield 12,172	18.07
Renalba 4117 - -	17.04½
Princess Mostar 9700	17.03
Bellini's Maid 15,170 -	15.01½
Bellini's La Biche 15,091	14.14½
Magnibel 7976 -	14.12
Susie La Biche 3d 15,171	14.06½
Lillian Mostar 10,364	14.03
Elmora Mostar 15,955 -	14.00

PILOT, JR., 141.
Son of Pilot 3 and Jenny 287.
1st sire of

Silver Rose 4753	16.14

2d sire of

Renalba 4117 -	17.04½
Romp Ogden 2d 4764 -	15.05
*Woodland Margaret 6215	14.10½
Opaline 7590 -	14.10
Pixie 4115	14.00

3d sire of

Roonan 5133	18.12
Leoni 11,868 -	18.02
Bonnie Yost 7943 -	18.02
Dudu of Linwood 8336	16.15
Kate Gordon 8387	15.15
Jaquenetta 10,958	14.06
Gem of Sassafras 8434	14.03½
Lucetta 6856 -	14.03

4th sire of

Wakena 19,721	16.00
Dora Doon 12,909	15.00
Fall Leaf 8587	14.08
Adora 18,569 -	14.03

PINE CLIFF 1106.
Son of Cliff 176 and St. Catherine 408.
1st sire of

St. Perpetua 2d 5557	14.00

2d sire of

Ramilly 17,075 -	14.00

PIONEER 368.
Son of Potomac 153 and Candour 325.
4th sire of

Armon 10,862	16.13½
Azelda 3d 7022	15.02

PISTOL 53.
Son of Bill 50 and Fanny 44.
3d sire of

Mirtha 3437	17.13½
Mirth's Blanche 19,572	17.13½

4th sire of

Lady Josephine 11,560 (8 days)	19.02
Merry Burlington 7600 -	15 04

PLANET 1130.
Son of Mack 722 and Maiden 1464.
3d sire of

Armon 10,862	16.13½

PLUTO 232.
Son of Lindo 233 and Primrose 562.
3d sire of

Mirth's Blanche 19,572	17.13½

POPE 738.
Son of Brunx Bashaw 145, and Jude 1030.
3d sire of

Nellie Gray of Clermont 10,905	14.01

POTOMAC 153.
Son of Comus 54 and Clara 148.
1st sire of

Nellie 1507	14.02

2d sire of

Beulah of Baltimore 3270	14.06½

3d sire of

Cyrene 4th 480	17.01
Ochra 2d 11,516 -	16.06½
Jessie Lee of Labyrinth 5290	14.07
Buttery 3502 -	14.01
Miami Prize 8100	14.00

4th sire of

Conover's Beauty 12,650	18.00
Cerita of Meadowbrook 5056	17.08
Maudine of Elmwood 8718	16.15
Pattie Mc. 3d 4754 - -	16.08
Lily of Maple Grove 5079	16.03
Glendelia 10,524	15.12
Fancy Juno 6086	15.10
Oitz 8649 - -	15.01
Alice of the Meadows 20,748	14.12
Gold Princess 8809	14.12
Litza 6338	14.03
Buttery 3502	14.01
Pixie 4115	14.00

PREMIUM 7.
3d sire of

Princess of Mansfiel d 8070	15.02
New London Gipsey 11,667	14.08
St. Nick's Flora 16,195	14.00

4th sire of

Attractive Maid 16,925 -	16.13
Pierrot's Lady Bacon 12,482	16.10

PRIDE OF THE ISLAND 5416.
1st sire of

Lenoreisa 16,233 -	21.00
Island Dots 17,003	14.09

PRINCE 55.
Son of Commodore 56 and Duchess 82.
1st sire of

Grace 2d 919	20.00
Oxalis 606 - - -	15.00

2d sire of

Reckless 3569	17.08
Haddie 921 -	16.00
*Belle of Inda 3867	15.01½
Oxalis 2d 15,631	15.00
Memento 1913	14.05

3d sire of

Mirtha 3437 - - -	17.13½
Mirth's Blanche 19,592	17.13½
Rosa of Glenmore 3179	17.10
Embla 4799	17.08
Reckless 3569	17.08
*Dora O. 11,703 -	17.03
Allie Minka 2982 -	14.06½
Maggie May 2d 12,926	14.06
Memento 1913	14.05
Cigarette 2849	14.04
Muezzin 3670 -	14.00
Naomi Cramer 8628	14.00

4th sire of

Mary M. Allison 6308	20.14
*Miskwa 15,472 -	19.06
Bonnie Yost 7943 -	18.02
Mirth's Blanche 19,592 -	17.13½
Florinanna 9862	17.05
*Dora O. 11,703 -	17.03
Urbana 5597 -	16.00
Merry Burlington 7600	15.04
Dark Cloud 9364	15.03½
Allie Minka 2982 -	14.06½
Rose of Rose Lawn 9365	14.06
Cigarette 2849 -	14.04
Ada Minka 15,562	14.00

PRINCE 120.
Son of Dick 118 and Nellie Bly 226.
4th sire of

Lady Alice of Hillcrest 7450	16.08
Gray Therese 5322	16.00

PRINCE 155 J. H. B.
3d sire of

Royal Princess 2d 12,346	17.12

PRINCE 462.
Son of Splendid 2 and Hebe 183.
3d sire of

Dimple 3248	16.11

4th sire of

Orphean 4636 -	15.07
Gem of Sassafras 3434 -	14.03½

PRINCE ALBERT 26.
Sire on I. of Jersey, dam Jewess 219.
4th sire of

Countess of Scarsdale 18,633	14.06

PRINCE ALBERT 119.
2d sire of

Countess of Warren 3896	14.00

3d sire of

Jo 5th 280 - -	17.08
Warren's Duchess 4622 -	16.01
Monmouth Duchess 3895	14.07

4th sire of

Dot of Bear Lake 6170	19.04
Ida of Bear Lake 6169	16.00
Mary of Bear Lake 6171 -	14.07

PRINCE CHARLES 816.
Son of Prince 55 and Clara 148.
3d sire of

Mary M. Allison 6308	20.14
*Miskwa 15,472	19.06
*Dora O. 11,703	17.03
Dark Cloud 9364 -	15.03½
Rose of Rose Lawn 9365	14.06

PRINCE JOHN 22.
Son of St. Clement 10, dam imported.
2d sire of

*Belle of Bloomfield 4331	14.00

3d sire of

White Clover Leaf 4512	17.15
Rosa Miller 4333	17.07
Nellie 1507 -	14.02
Lady Brown 433	14.00
*Sutliff's Rosy 4104	14.00

4th sire of

*Optima 6715	23.11
Hepsy 2d 12,008	17.08
Rosa Miller 4333	17.07
Cyrene 4th 480	17.01
Silver Rose 4753	16.14
Myrtle 2d 211 -	15.12
Romp Ogden 2d 4764	15.05
Arietta 5115 -	15.00
Lady Brown 4th 6911	14.12
Cowle's Nonesuch 6199	14.12
Bloomfield Lady 6912	14.12
Beulah of Baltimore 3270	14.06½
Lobelia 2d 6650 -	14.06
Beauty Bismarck 4967 -	14.01
Jessie Leavenworth 8248	14.00
*Churchill's Betsey 4105	14.00

PRINCE OF CROTON 2490.
Son of Tocsin 1912 and Glendelia 4436.
1st sire of

Princes' Bloom 9729 -	14.03

2d sire of

Grinnelle Lass 11,859	16.10

PRINCE OF JERSEY 66.
3d sire of

Bounty 1606 - -	14.00

4th sire of

Lady Mel 2d 1795	21.00
Thisbe 2d 2201	19.01½
Jo 5th 280 - -	17.08
Champion Chloe 12,255	15.05½
Bounty 1606	14.00
Lady Brown 433	14.00
*Bonmari 7019 -	14.00
*Allen's Fawnette 3722 -	14.00
*Brightness 2211 -	14.00
Silver Belle 4313 -	14.00

PRINCE OF ROSE CROIX 1893.
Son of Fairfax 530 and Minnie 398.
3d sire of

*Wabash Girl 14,550	16.00

PRINCE OF ORANGE 184.
Son of Saturn 94 and Daisy 440.
3d sire of

Mamie Coburn 3798	17.08
Pet Anna 1608	14.00

4th sire of

Welma 5942 -	17.08
Elinor Wells 12,068	14.00

PRINCE OF SIDNEY 2665.
Son of Noble 901 and Dahlia of Babylon 2346.
1st sire of

Daisy of Chenango 18,582	14.07

PRINCE OF THE VALLEY 88 J. H. B.
Son of Welcome 166 J. H. B. and Queen of the Valley 740 J. H. B.
1st sire of

Faith of Oaklands 19,696	17.04

PRINCE OF WALES, I. OF J.
Son of " Old Noble" and Duchess 24 J. H. B.
2d sire of

Lisette 483 J. H. B	16.04
Jeanne Le Bas 2476	.15.08
Fille de l'Air 2474	14.00

3d sire of

Nelly 6456 -	21.00
Garenne 1575 J. H. B.	17.08
Fear Not 6059 - -	17.03
Dairy Pride 4th 521 J. H. B.	16.00
Dairy Pride 6th 21681	16.00
Atricia 6029 -	15.03
Queen of Ashantee 14,554	15.02
Regina 2d 2475	14.08
Esperanza	14.03

4th sire of

Nancy Lee 7618 -	26.08½
Chrome Skin 7881 -	20.10
Queen of Delaware 17,029	18.13
Butter Star 7799	18.04½
Regina 4th 12,732	17.13½
Faultless 12,018 -	17.05½
Faith of Oaklands 19,696	17.04
Fear Not 6059 -	17.03
Cream of Sidney 17,028	17.02½
Fear Not 2d 6061 -	16.02
Lucilla Kent 8892 -	15.10
Maid of Five Oaks 7178	15.04
Fan of Grouville 7458	15.00
Walkyrie 5708	14.01

PRINCE OF WARREN 1512
Son of Southampton 117 and Golddrop 222.
1st sire of

Dot of Bear Lake 6170 -	19.04
Conover's Beauty 12,650	18.00
Ida of Bear Lake 6169 - -	16.00

Glendelia 10,524 - -		15.12
Mary of Bear Lake 6171		14.07

2d sire of

Countess Lowndes 26,874		17.08
Witch Hazel 4th 6131		15.05½

PRIOR 249.
Son of Commodore 56 and Milly 526.
1st sire of

Heartsease 503		15.00

2d sire of

Pansy 602	-	14.00

3d sire of

Reckless 3569	-	17.08
Helve 4565		14.00

4th sire of

Joan d'Arc 2163 -	-	16.13½
Willis 2d 4461	-	16.03
Lebanon Daughter 6106		14.04
Lebanon Lass 6108 -		14.02

PRIZE DUKE 942.
Son of Clive 319 and Jersey Prize 1267.
2d sire of

Fancy Juno 6086	15.10
Oitz 8649	15.01

3d sire of

Lady Cloud 19,358	16.10

PROGRESS 286 J. H. B.
1st sire of

Beauty Romereil 26,090	18.09

PROXY 1714.
Son of Pertinatti 713 and Roxana 1761.
2d sire of

Fall Leaf 8587	-	14.08
Adora 18,569	-	14.03

PULASKI 1932.
Son of John Allen 1494 and Lady Orton 2667.

Cowle's Nonesuch 6199	14.12

PURITANI 2275.
Son of Barney 1491 and Milkmaid 2d 5176.
1st sire of

Jeannette Montgomery 5177	20.00

QUAKER 887.
Sire on I. of Jersey, dam Queen of Staatsburg 2234.
3d sire of

Siloam 17,623 -		18.10
Almah of Oakland 11,102	-	16.14
Kate Gordon 8387 -		15.15
Lady Adam 2d 6529	-	15.03
*Myra Overall 10,317		14.00

QUOGUE 690.
Son of Monadnock 258 and Frisky 1470.
1st sire of

Lady Penn 5314	16.00

2d sire of

Goldthread 4945 -	16.09

RAGHORN 175.
Son of Splendid 2 and Rose 394.
1st sire of

Lady Ives 1708	18.00

2d sire of

Lady Ives 3d 6740 -	14.03
*Belle of Bloomfield 4331	14.00

3d sire of

Rosa Miller 4333	17.07
Louvie 3d 6159	14.13
Kate Daisy 8204	14.04

4th sire of

Rosa Miller 4333	17.07
Percie 14,937	16.12
Albena 15,995	16.08
Œnone 8614 -	15.14
Bloomfield Lady 6912	14.12

RAJAH 340.
1st sire of

Oonan 1485 - -	22.02½
Fautnie 1271	15.06
Spirea 3915 -	14.00
Coral 2707	14.00

2d sire of

Hazen's Nora 4791	20.04
Roonan 5133 -	18.12
Mamie Coburn 3798	17.08
Odelle Sales 15,564	16.03
Callie Nan 7959 -	16.02
Arawana Buttercup 6052	15.05
Arawana Poppy 6053	15.02
Bellini's Maid 15,170	15.01½
*Belle of Inda 3867 -	15.01½
Bellini's La Biche 15,091	14.14½
Pet Lee 7993 - -	14.12
Pride of the Hill 4877	14.08
Enid 2d 10,782 -	14.07½
*Belle of Milford 7445	14.07
Susie La Biche 3d 15,171 -	14.06½
Rose of Hillside 3866 -	14.03½
Maggie May 3255 -	14.02
Myrtle of Ridgewood 7858	14.01
Gilt Edge 2d 4420	14.00
*Ramily 17,075 -	14.00

3d sire of

Dudu of Linwood 8336	16.15
Lucky Belle 6037	16.14
Arawana Queen 5368 -	16.09
Julia Evelyn 6007	15.15½
Valerie 6044 -	15.13
Calypris 5943 -	15.0
Duchess Caroline 3d 6041	15.08
Bettie Dixon 4527 -	15.00
Cosetta 15,991	14.11
Coronilla 8867 - -	14 09½
Pride of the Hill 4877	14.08
Maggie May 2d 12,926	14.06
Maggie C. 12,216 -	14.06
Walkyrie 5708	14.01

Starkville Beauty 4897	-	14.00
Vivalia 12,760 -	-	14.00

4th sire of

Atlanta's Beauty 12,949		21.03
Calendine 9415		17.09
Armon 10,862	-	16.13½
Favorite's Rajah Rex 16,153		15.00
Dora Doon 12,909 -		15.00
Mountain Lass 12,921		14.09
Gilt Edge C. 12,223 -		14.03½
Minnie Lee 2d 12,941	-	14.03
Therese M. 8364		14.02
Vivalia 12,760		14.00
La Rosa 10,078	-	14.00

RALPH 957.
Son of St. Helier 45 and Ibi 671.
1st sire of

Mhoon Lady 6560 -	-	17.03
Cenie Wallace 2d 6557		15.04½
Florry Keep 6556		14.14

RALPH GUILD 1917.
Son of Planet 1130 and Myce 18.10.
2d sire of

Armon 10,862	16.13½

RAMBLER OF ST. LAMBERT 5285.
Son of Stoke Pogis 3d 2238 and Bessy of St. Lambert 5482.
1st sire of

Rose of St. Lambert 20,426	20.03½
Rioter's Ruth 14,882	14.12
Rioter's Beauty 14,894	14.00

RAMCHUNDER 718.
Son of Rajah 340 and Nelly 55.
1st sire of

Mamie Coburn 3798	-	17.08
Belle of Inda 3867 -	-	15.01½
Rose of Hillside 3866		14.03½

3d sire of

La Rosa 10,078	-	14.00

RANGER 1231.
Son of Rival 395 and Rachel 1056.
1st sire of

Mhoon Lady 6560	17.03
Cenie Wallace 2d 6557	15.04½
Florry Keep 6556	14.14

RATTLER 20 J. H. B.
2d sire of

Regina 4th 12,732	-	17.13½
Faultless 12,018		17.05½

3d sire of

Fillpail 16,530	15.11

RECTOR 1458.
Son of Pertinatti 713 and Roxana 2d 2532.
1st sire of

Bonnie Yost 7943	-	18.02
Leoni 11,868		18.02

| Dudu of Linwood 8336 | 16.15 |
| Lucetta 6856 -   - | 14.08 |

2d sire of

| Dora Doon 12,909 | 15.00 |

RED CLOUD 529.

Sire on I. of Jersey, dam Corona 1159.
2d sire of

| Fancy Juno 6086 | 15.10 |

3d sire of

Judith Coleman 11,391 -	17.05
Lady Cloud 19,358	16.10
Aleph Judea 11,389	15.01¾

RED CLOUD 2D 2260.

Son of Red Cloud 529 and Famosa 1364.
1st sire of

| Fancy Juno 6086 | 15.10 |

2d sire of

Judith Coleman 11,391 -	17.05
Lady Cloud 19,358	16.10
Aleph Judea 11,389	15.01½

RED KNIGHT 666.

Son of Orange Peel 129 J. H. B. and Bergerette 3711 (303 J. H. B.)
2d sire of

| Charmer 4771 | 14.12 |

3d sire of

| Lady Josephine 11,560 (8 days) | 19.02 |
| Nelida 2d 8227   -   - | 15.02½ |

REEFER 209.

Son of Commodore 56 and Gazelle 187.
4th sire of

| *Dora O. 11,703 | 17.03 |

REINDEER 1194.

Son of Waterbury 1155 and Snowflakes 1004.
2d sire of

| Hepsy 2d 12,008   - | 17.08 |
| Jessie Leavenworth 8248 | 14.00 |

3d sire of

| Kitty Potter 9893 | 18.05 |
| Bell Rex 11,700 | 14.10 |

RELIEF 150.

Sire on I. of Jersey, dam Charity 320.
4th sire of

| Alice of Salem 5053 | .14.08 |

REMARKABLE 229 J. H. B.

Son of Orange Peel 2d 36 J. H. B.
1st sire of

Rosa of Belle Vue 6954 -	18.07½
Mary Jane of Belle Vue 6956	17.07
Caroline 12,019	14.08

2d sire of

| Countess Gasela 9571   - | 15.11 |

RENO 563.

Son of Rajah 340 and Renella 62.
2d sire of

| Calypris 5943 | 15.09½ |

REVENUE 1744.

Son of Ben Rajah 795 and Regina 2d 2475.
1st sire of

| Calendine 9415 | 17.09 |
| Cosetta 15,991 | 14.11 |

REWARD 190.

Son of Commodore 56 and Faith 78.
2d sire of

| Eureka McHenry 8841 | 14.00 |

REX 71 J. H. B.

Son of Orange Skin 19 J. H. B. and Regina 32 J. H. B.
1st sire of

| *Sultane 2d 11,373 | 23.08 |
| Merry Duchess 13,693 | 18.09½ |

REX 1330.

Son of Colt, Jr., 825 and Couch's Lily 3237.
1st sire of

Arawana Queen 5368 -	16.09
Princess Bellworth 6801 -	15.10½
Usilda 2d 6137 -	15.02½
Favorite's Rajah Rex 16,153	15.00
Louvie 3d 6159	14.13
Bell Rex 11,700	14.10
Princess Rose 6249	14.08
Lottie Rex 18,757	14.04
Jeannie Platt 6005	14.04
Pet Rex 20,166	14.02½
Hepsy 2d 12,008	14.00

2d sire of

Rosy Kate's Rex 13,192	18.08
Maggie Rex 28,623	17.00½
Elsie Lane 13,302	15.12

3d sire of

| Genevieve Sinclair 11,167 | 16.02 |

RHODES' BULL 824.

Sire one of J. T. Norton's bulls, dam Stanley Cow 2053.
2d sire of

| Maggie 3d 3221 | 17.08 |

3d sire of

| Maggie Rex 28,263 | 17.00½ |
| Chloe Beach 3981 | 14.08 |

4th sire of

Hepsy 2d 12,008 -	17.08
Arawana Queen 5368	16.09
Princess Bellworth 6801	15.10½
Arawana Buttercup 6052 -	15.05
Usilda 2d 6157 -	15.02½
Favorite's Rajah Rex 16,153	15.00
Louvie 3d 6159 -	14.13
Bell Rex 11,700	14.10
Princess Rose 6249	14.08
Chloe Beach 3931	14.08
Lottie Rex 19,757	14.04

Jeannie Platt 6005 -	14.04
Pet Rex 20,166	14.02½

RICHMOND 19.
Son of Splendid 2 and Meg 838.
4th sire of

Mollie Garfield 12,172	18.07
Lara 4306 - - -	17.08
Renalba 4117 -	17.04½
Bellini's Maid 15,170 -	15.01½
Bellini's La Biche 15,091	14.14½
Magnibel 7976 - -	14.12
Susie La Biche 3d 15,171	14.06½
Adina 1942	14.04

RIDGELY OF HAMPTON 3046.
Son of Orange Skin 1216 and Button 2d 3160.
1st sire of

Naomi's Pride 16,745	15.02

RINGLEADER 392.
1st sire of

Miss Blanche 2515 (10 days)	20.09

3d sire of

*Queen of Maple Dale Farm 7036 -	14.14
Village Maid 7069	14.00

4th sire of

Alphea Star 16,532 -	14.04½

RIOTER 670.
1st sire of

Duchess of Bloomfield 3653	20.00½
Su Lu 4705	17.15
Letitia 3977 . -	15.05
Lady Bloomfield 4704	14.12½

2d sire of

*Variella 6337 -	18.03¾
Princess Mostar 9700	17.03
Lady Bloomfield 4704	14.12½
Princess Bowen 9699	14.12
Lorella 12,913	14.07
Lucetta 6856	14.03

3d sire of

Jaquenetta 10,958	14.06
Elmora Mostar 15,955	14.00

RIOTER 746 E. H. B.
Son of Pedlar 631 E. H. B. and Rita.
2d sire of

Eurotas 2454 -	22.07
*Pet Gifford 3377	18.04½
Torfrida 3596	17.06½
Hebe 3d 3613 -	15.00
Rioter 2d's Venus 3658 -	14.00

3d sire of

Colie 8309 -	18.04
*Lanice 13,656 -	17.08
*Blossom of Hanover 13,655	17.08
Pyrrha 6100 -	16.14½
Typha 5870 -	16.11
Euphrates 9778 -	16.05

4th sire of

Mary Anne of St. L. 9770 - -	36 12½
Bomba 10,330 -	21.11½
La Belle Petite 5472 -	17.08
Diana of St. L. 6636	16.08
Maggie of St. L. 9776	16.03
Moth of St. L. 9775	16.02
Euphonia 6783 -	16.00½
*Pedro Alphea 13,889	15.05
Nazli 10,327 -	15.03½
Nimble 22,335	14.10
Smoky 13,733 -	14.09
Cupid of Lee Farm 5997	14.06
*Lady Golddust 2d 19,861	14.01

RIOTER 2D 469.
Son of Rioter 746 E. H. B. and "No. 78,"
Dauncy's Sale.
1st sire of

Eurotas 2554 -	22.07
*Pet Gilford 3377	18.04½
Torfrida 3596 .	17.06½
Hebe 3d 3613 -	15.00
Rioter 2d's Venus 3658 -	14.00

2d sire of

Colie 8309 - -	18.04
*Blossom of Hanover 13,655	17.08
*Lanice 13,656 -	17.08
Pyrrha 6100 -	16.14½
Typha 5870	16.11
Euphrates 9778 -	16.05

3d sire of

Bomba 10,330	21.11½
Euphonia 6783 -	16.00½
*Pedro Alphea 13,989	15.05
Nazli 10,327 -	15.03½
Nimble 22,335	14.10
Smoky 13,733 -	14.09
*Lady Golddust 2d 19,861	14.01

ROANOKE 1448.
Son of Beechnut 109 and Princess 366.
1st sire of

Gold Lace 10,726	14.13

2d sire of

Gold Mark 10,727	14.14

RIVAL 395.
Sire on I. of Jersey, dam Rosamond 1053.
2d sire of

Village Maid 7069	14.00

4th sire of

Lydia of Libby 1698	15.03

ROB 874.
Son of Ringleader 392 and Elsie 1054.
2d sire of

*Queen of Maple Dale Farm 7036	14.14

3d sire of

Alphea Star 16,532	14.04½

ROBBINS 953.
Son of Albert 44 and Victoria 2d 419.
1st sire of

Dusky 2525	16.10
Pretty 2526	14.00

2d sire of
*Kitty Lake 8250   15.08½
3d sire of
Cordelia Baker 8814 -   17.09
Colt's La Biche 6399   -   17.02½
Phyllis of Hillcrest 9067   14.12
Roll of Honor 13,610   14.12
4th sire of
Alfieda 6744   -   -   16.04
Susie La Biche 3d 15,171   14.06½
Hurrah Pansy 12,153   14.01½

ROB ROY 17.
1st sire of
Belle Hartford 7718   15.00
Eugenie 2d 1623 -   14.00
2d sire of
Belle Grinnelle 4073   18.08
White Clover Leaf 4512   17.15
Fair Starlight 1745 -   17.07½
Corolla 4392   -   -   --   17.04
*Belle of Saybrook 6875   15.01½
Jersey Cream 2d 8519   14.12
Princess Rose 6249 -   14.08
Chloe Beach 3931   -   14.06½
Lucy Gaines' Buttercup 5058   14.00
St. Perpetua 2d 5557 ·-   14.00
3d sire of
Countess Potoka 7496   18.15
Hepsy 2d 12,008   17.08
Katie Bashford 15,982   17.00
Arawana Queen 5368 -   16.09
Lily of Maple Grove 5079   16.03
Princess Bellworth 6801   15.10½
Arawana Buttercup 6052   15.05
Usilda 2d 6157   -   -   15.02½
Favorite's Rajah Rex 16,153   15.00
Louvie 3d 6159   -   14.13
Bell Rex 11,700   14.10
Princess Rose 6249   -   14.08
Jeannie of the Vale 9553   14.06½
Jeannie Platt 6005   14.04
Lottie Rex 18,757   14.04
Pet Rex 20,166 -   -   14.02½
Belle Grinnelle 3d 16,503   14.02
4th sire of
Hillside Gem 16,640   20.00
Gardinier's Ripple 11,693   19.12½
Rosy Kate's Rex 13,192   18.08
Maggie Rex 28,263   17.00½
Grinnelle Lass 11,859   16.10
Lida Mullin 9198   -   16.08
Olie's Lady Teazle 12,307   16.05
Lady Cecelia 24,821   16.01
Elsie Lane 13,302   15.12
Euphorbia 11,229   14.09½
Princes' Bloom 9729   14.08
Lizzie D. 10,408   :   14.00

RODERICK 128.
Son of Blucher 48 and Lady Bowen 354.
4th sire of
Yellow Locust 10,679   14.10½

ROGER 121.
Son of Jupiter 122 and Jessie 1006.
1st sire of
ʋ 5th 280   17.08

3d sire of
Lady Alice of Hillcrest 7450   16.03
Gray Therese 5322   16.00
Oitz 8649   15.01

ROLAND OF DEERFOOT 8363.
Son of Albion 490 and Rose 2714.
2d sire of
Roland's Bonnie 2d 18,054   19.02

ROLLA AZULINE 804.
2d sire of
Starkville Beauty 4897   ,   14.00
3d sire of
Mountain Lass 12,921   14.09
Minnie Lee 2d 12,941   14.03

ROMEO 982.
Son of Living Storm 173 and Nellie 2d 2577.
3d sire of
Cordelia Baker 8814   17.09
Mary Clover 9998   14.15
4th sire of
Kate Daisy 8284   14.04

ROMEO OF NEEDHAM 8361.
Son of Abraham 228 and Victoria 554.
4th sire of
Roland's Bonnie 2d 18,054 -   19.02

ROMULUS 181 J. H. B.
Son of Hero 90 J. H. B. and Stella 705 J H. B.
1st sire of
Bergerelia 15,546   14.01½

ROSLAND 1553
Son of Prince Charles 816 and Belle 194.
2d sire of
*Miskwa 15,472   19.06

ROSSMAN 1128.
Son of Pilot, Jr., 141 and Rosa 122
2d sire of
Gem of Sassafras 8434   14.03½

ROXBURY 247.
Son of Commodore 229 and Rose 709
1st sire of
Angela 1682   -   14.02
2d sire of
Duchess of Bloomfield 3653   20 00½
Su Lu 4705   17.15
Vixen 7591 -   17.06
Pattie Mc. 3d 4754   16.08
*Pulsatilla 7551   16.03
Letitia 3977   15.05
Bathsheba 2556   -   14.01

3d sire of

Roonan 5133	18.12¾
Variella 6337	18.03
Lara 4306 -	17.08
Pattie Mc. 3d 4754	16.08
Urbana 5597 -	16.00
Kate Gordon 8387	15.15
Lorella 12,913	14.07
Lucetta 6856 -	14.03
Litza 6388 -	14.03
Variella of Linwood 10,954	14.01
Pixie 4115 - -	14.00

4th sire of

Jeannette Montgomery 5177	20.00
Bonnie Yost 7943	18.02
Leoni 11,868 - -	18.02
Dudu of Linwood 8336	16.15
Urbana 5597 -	16.00
Silenta 17,685	15.10
Fall Leaf 8587	14.08
Epigea 4631 -	14.07
Jaquenetta 10,958	14.06
Adora 18,569	14.03
Lucetta 6856 -	14.03
Silene 4307	14.00
Jule 3640	14.00

ROYAL ROB 598.

Sire on I. of Jersey, dam Lilly of Ipswich 597.

4th sire of

Countess of Scarsdale 18,633	14.06

RUMSON 741.

Son of Narragansett 536 and English Beauty 449.

1st sire of

Nellie Maitland 4450	16.00

RUPERT 1456.

Son of Pertinatti 713 and Roxana 1761.

1st sire of

Roonan 5133	18.12

RUSS'S SAM 1709.

Son of Waterbury 1155 and Dewdrop 4101.

1st sire of

*Sutliff's Rosy 4104 -	14.00

2d sire of

*Churchill's Betsey 4105	14.00

3d sire of

Webster's Pet 4103 -	14.02

SAILOR 169.

2d sire of

Jersey 3260 -	15.06
Topsey Roxbury 7796	15.00

4th sire of

Maud Lee 2416	23.00

SALADIN 447.

Son of Cadmus 4 and Woodbine 517.

1st sire of

Rosa of Glenmore 3179 -	17.10
Embla 4799	17.08

2d sire of

Naomi Cramer 8628	14.00

SAM 402.

Son of Comus 54 and Diana 77.

2d sire of

Allie Minka 2982	14.06½

3d sire of

Valma Hoffman 4500 -	21.09
Naomi's Pride 16,745	15.02
Bessie Ridgely 8293 -	14.12
Celia Belle 5865 -	14.03
Ada Minka 15,562	14.00

SAM 980

Sire on I. of Jersey, dam Eugenie 792.

1st sire of

Sal Soda 3721	14.07

2d sire of

Countess Potoka 7496 -	18.15
Lady Gray of Hill Top 6850	18.12
Summerline 8001 -	18.06
Silveretta 6852	16.09

3d sire of

Silveretta 6852 - -	16.09
Lady Gray of Hill Top 2d 14,641	14.12
Lady Gray of Hill Top 3d 14,642 -	14.02

SAMSON 1079.

Son of Waterpower 756 and Ocean Queen 1405.

1st sire of

Merlette 4988	16.00

SAM WELLER 40.

Son of Iron Duke 18 and Kitty Clyde 30.

3d sire of

Hazen's Nora 4791 -	20.04
Arawana Buttercup 6052	15.05
Arawana Poppy 6053	15.02
Belle of Milford 7445	14.07

4th sire of

Arawana Queen 5368	16.09
Coronilla 8367 -	14.09½
Maggie C. 12,216 -	14.06
Maggie May 2d 12,926	14.06

SAM WELLER 271.

2d sire of

Hennie 3335	15.00

3d sire of

Masena 25,732	20.07
Hilda A. 2d 11,120	20.00
Hilda D. 6683 -	18.05
Lily of Burr Oaks 11,001	15.13
Niva 7523 - -	15.08
Nibbette 11,625	14.07
Nordheim Creamer 9758	14.00

4th sire of

Jersey Queen of Barnet 4201 A. H. B.	19.12
Roland's Bonnie 2d 18,054	19.02
Snowdrop F. W. 16,948	14.08

SAM WELLER, JR., 1352.
Son of Sam Weller 271 and Nora 956.
2d sire of

Masena 25,732	20.07

4th sire of

La Pera 2d 13,404	14.08

SANCHO BOY 1576.
Son of Jersey Boy 272 and Rose 2714.
3d sire of

Countess Queen 13,519	18.03

SANS PEUR 201 J. H. B.
Son of Welcome 166 J. H. B. and Fanchon 1432 J. H. B.
1st sire of

Fear Not 6059	17.03
Buttercup 17,285	16.08
Fan of Grouville 7458	15.00

2d sire of

Fear Not 2d 6061	16.02
Lucilla Kent 8892	15.10

SANTA ANA 221.
Sire on I. of Jersey, dam Dolly 545.
3d sire of

Maud Lee 2416	23.00
Gold Lace 10,726	14.13

4th sire of

Maud Lee 2416	23.00
Gold Mark 10,727	14.14
Alice of the Meadows 20,748	14.12

SANTA CLAUS 30.
Son of Jack Frost 31 and White Heart 63.
2d sire of

Princess of Mansfield 8070	15.02
New London Gipsey 11,667	14.08

3d sire of

Peggy Leah 3097	18.12
Pierrot's Lady Bacon 12,482	16.10
Jessie Leavenworth 8248	14.00

4th sire of

Dimple 3248	16.11
Myrtle 2d 211	15.12
Rene Ogden 1568	15.00
Mary Clover 9998	14.15
Bell Rex 11,700	14.10
Litty 8017	14.00

SAPPER 1026.
Son of Partisan 235 and Silver Gray 1342.
1st sire of

Mirth's Blanche 19,572	17.13½

SCION 1033.
Son of Red Knight 666 and Cybele 3d 1270.
1st sire of

Charmer 4771	14.12

2d sire of

Lady Josephine 11,560 (8 days)	19.02
Nelida 2d 8227	15.02½

4th sire of

Countess Scarsdale 18,633	14.06

SCOTIA 1154.
Son of Litchfield 674 and Rosebud 4th 477.
2d sire of

Albena 15,995	16.03

SCROOGE 369.
Son of Sam Weller 271 and Susie 959.
2d sire of

Niva 7523	15.08

SARATOGA 135.
2d sire of

Oitz 8649	15.01
Silver Belle 4313	14.00
Gilt 4th 4208	14.00

3d sire of

Estrella 2831	14.12
Gilt Edge 2d 4420	14.00

4th sire of

Lady Louise 4339	15.00
Gilt Edge C. 12,223	14.03½

SARATOGA 2D 1272.
Son of Saratoga 135 and Dainty 2d 248.
4th sire of

Atlanta's Beauty 12,949	21.03

SARK 123.
Son of Prince of Jersey 66 and Belle 225.
2d sire of

Bounty 1606	14.00

3d sire of

Thisbe 2d 2201	19.01½
Jo 5th 280	17.08
*Bonmari 7019	14.00
Silver Belle 4313	14.00

4th sire of

Ida Bashan 4725	18.00
Forget-Me-Not-O. 10,564	15.04
*Belle of Inda 3867	15.01½
Estrella 2831	14.12
Irene of Short Hills 5137	14.06½
Pet Anna 1608	14.00
St. Perpetua 2d 5557	14.00
Vivalia 12,760	14.00

SARPEDON 930.
Son of Mercury 432 and Europa 176.
2d sire of

Bomba 10,330	21.11½
Nazli 10,327	15.08½
Robinette 7114	14.01
Lady Golddust 2d 19,861	14.01

3d sire of

*Pedro Alphea 18,889	15.05

SATURN 94.
1st sire of

*Alphea 171	24.08

2d sire of	
*Pet Gilford 3317	18.04½
Europa 176	15.00
*Leda 799	14.05½

3d sire of	
Eurotas 2454	22.07
*Locusta 5143	21.07
Phædra 2561	19.13
Nymphæa 5141	18.07½
Oak Leaf 4769 -	17.10
Torfrida 3596	17.06½
Lass Edith 6290	17.00
Euphrates 9778 -	16.05
Gray Therese 5322	16.00
Lerna 3634	15.12
Zalma 8778 -	15.05
Purest 13,730 -	15.04
Clytemnestra 2455	15.03½
Iola 4627 -	15.02½
Marvel 13,734	15.01
Trudie 2d 4084	15.00
Maple Leaf 4768	14.12
Nimble 22,335	14.10
Smoky 13,733	14.09
Ideal Alphea 18,755	14.06
Renown 13,729	14.06
Richness 16,536	14.06
*Leda 799 -	14.05½
Alphea Star 16,532	14.04½
Silversides 3857	14.03
Alphetta 16,531	14.02½
Vestina 2458	14.02
Lernella 22,322 -	14.01½
Ballet Girl 18,750 -	14.01
Alphea Jewel 22,331	14.00
Edith 4th 817	14.00
Spring Leaf 5796	14.00

4th sire of	
*Katie Kohlman 7270	23.10
Phædra 2561	19.13
Nymphæa 5141	18.07½
Colie 8309	18.04
Zampa 2194 - -	18.00
Zittella 2d 11,922	17.08½
*Lanice 13,656 -	17.08
*Blossom of Hanover 13,655 -	17.08
Mamie Coburn 3798 -	17.08
Lass Edith 6290	17.00
Typha 5870 - -	16.11
Lady Alice of Hillcrest 7450	16.03
Euphonia 6783 -	16.00½
Bessie S. 5002	16.00
Myra 2d 6289 -	16.00
Pride of Corisande 5323	16.00
Gray Therese 5322	16.00
Clytemnestra 2d 5868	16.00
Fillpail 16,530	15.11
Malope 2d 11,923	15.10
Idalene 11,841	15.08½
Crust 4775 -	15.07
Forget-Me-Not-O. 10,564	15.04
Purest 13,730 -	15.04
Clytemnestra 2455	15.03½
Trudie 2d 4084	15.00
Hebe 3d 3613	15.00
Faustine 10,354	14.14½
Ideal 11,824	14.12½
Estrella 2831 -	14.12
Smoky 13,733	14.09
Hartwick Belle 7722 -	14.08
Renown 13,729	14.06
Richness 16,536	14.06

Kate Daisy 8204 -	14.04
Bessie Bradford 7269	14.02
Lernella 22,322	14.01½
Robinette 7114 -	14.01
Honey Drop 10,033 -	14.00½
Pet Anna 1608	14.00
Rioter 2d's Venus 3658	14.00
St. Nick's Flora 16,195 -	14.00

SAUGATUCK 1144.

Son of Manfred 510 and Rose Standish 1865.

2d sire of	
Alfleda 6744 -	16.04
Nannie Fitch 9143 -	14.04

3d sire of	
Alfritha 13,673	15.03

SCHINCHON 1132.

Son of Pierrot 636 and Beauty 804.

1st sire of	
Polly Clover 7052	16.15

2d sire of	
Kitty Potter 9893	18.05

SECOND DUKE OF HILLCREST 1086.

Son of Standard 553 and Picture 1533.

2d sire of	
Roll of Honor 13,610	14.12

SECOND IRON DUKE 202.

Son of Iron Duke 18 and Niobe 99.

2d sire of	
Cerita of Meadowbrook 5056	17.08
Charmer 4771	14.12
Epigæa 4631	14.07

SEECONNET 1480.

Son of Miantonomoh 730 and Mamie of Newport 2013.

2d sire of	
Katie Bashford 15,982 - -	17.00

3d sire of	
Hillside Gem 16,640	20.00

SHAKER 21.

Son of Prince John 22 and Clover 20.

3d sire of	
Myrtle 2d 211 -	15.12

4th sire of	
Myrtle of Ridgewood 7858	14.01

SHARPSHOOTER 2406.

Son of Expounder 1148 and Symphony 1400.

1st sire of	
Alice Herrick 8787	14.14

SHARPSHOOTER OF ATLANTA 3011.
Son of Grand Duke Alexis 1040 and Zina
2d 3082.
1st sire of

Tenella 2d 19,521	28.12

SHELDON 5250.
Son of Lorne 5248 and Hebe of St. Lambert
5117.
1st sire of

Maggie Sheldon 23,588	-	15.03

SHIRLEY 1613.
Son of Mogul 532 and Niobe 4th 509.
1st sire of

Queen of Prospect 11,997	14.02

SIGNAL 1170.
Son of Marius 760 and Pansy Morris 2060.
1st sire of

*Optima 6715	23.11
Tenella 6712 -	22.01½
Croton Maid 5305	21.11½
Valhalla 5300	17.00
Belle of Patterson 5664	16.06
Œnone 8614	15.14
Edwina 6713 -	15.13
Fanny Taylor 6714	15.12
Signalana 7719	15.04
Aldarine 5801	15.01½

2d sire of

Fadette of Verna 3d 11,122	22.08½
Fairy of Verna 2d 10,973	20.03¾
Hilda A. 2d 11,120 -	20.00
Gardinier's Ripple 11,693	19.12½
Tenella 2d 19,521	18.12
Gazella 3d 9355 -	16.08
Genevieve Sinclair 11,167	16.02
Rupertina 10,409	15.12½
Reception 3d 11,025	15.08½
Euphorbia 11,229 -	14.09½
Lady Clarendon 3d 17,578	14.05½
Signetilia 16,833 -	14.03½
Sadie's Choice 7979 -	14.00

3d sire of

Atlanta's Beauty 12,949	21.03
Signetilia 16,833 -	14.08½

SIGNALDA 4027.
Son of Signal 1170 and Alda 3873.
1st sire of

Signetilia 16,833	14.03½

SILVERLOCKS 546.
3d sire of

Siloam 17,623	18.10
Silvia Baker 7893 -	16.04
Countess Coomassie 19,339	15.08½

SILVERLOCKS, JR., 699.
Son of Silverlocks 546, and Kathleen 1767.

2d sire of

Siloam 17,623	18.10
Silvia Baker 7873 -	16.04
Countess Coomassie 19,339	15.08½

4th sire of

Favorite's Rajah Rex 16,153	15.00

SILVER MINE 1658.
Son of Silverlocks, Jr., 699 and Minerva
1529.
1st sire of

Siloam 17,623 -	-	18.10
Silvia Baker 7893 -		16.04
Countess Coomassie 19,339 -		15.08½

SIMON PETER 1848.
Son of Pansy's Albert 1008 and Brightness
2211.
4th sire of

*Wabash Girl 14,550	16.00

SIR CHARLES 131
1st sire of

Violet 272	17.08
Carrie 3894	16.08
Julia 3893	16.08

2d sire of

Warren's Duchess 4622 -	16.01
My Queen 12,614 -	15.08
Monmouth Duchess 3895	14.07
Countess of Warren 3896	14.00

3d sire of

Dot of Bear Lake 6170	19.04
Warren's Duchess 4622 -	16.01
Ida of Bear Lake 6169	16.00
Mary of Bear Lake 6171	14.07

4th sire of

Dot of Bear Lake 6170	19.04
Countess Lowndes 26,874	17.08
Mary of Bear Lake 6171	14.07

SIR DAVY 84.
Sire on I. of Jersey, dam Jersey Maid 94
1st sire of

Beulah of Baltimore 3270	14.06½

2d sire of

Bessie Ridgely 8293 -	14.11½
Jessie Lee of Labyrinth 5290	14.07

3d sire of

Grace Davy 8292	14.02

4th sire of

Bessie Ridgely 8293	14.11½

SIR GEORGE OF ST. LAMBERT 6086.
Son of Stoke Pogis 3d 2238 and Pride of
Windsor 483.
1st sire of

Rioter's Nora 21,778	15.09

SIR JOHN 525.
Son of Paterson 11 and Daisy 2d 609.
4th sire of

Pattie Mc 3d 4754 -	16.08
Cowle's Nonesuch 6199 -	14.12

SIR SAMUEL CUNARD 2231.
Son of Scotia 1154 and Locust 3631.
1st sire of

Alhena 15,995	16.08

SISSON 23.
Son of Splendeus 16 and Sisson's Dam 22.
4th sire of

Canto 7194	15.12
*Mica 1983	15.12

SMOKE SMITH 4844.
Son of Brigand 1899 and Selika 1805.
1st sire of

Atlanta's Beauty 12,949	- 21.03

SMITH OF DARLINGTON 2458.
Sire on I. of Jersey, dam Premium of Darlington 5572.
1st sire of

Anna Smith 10,324 -	- 15.06
Nellie Darlington 5956	- 15.03

2d sire of

Bomba 10,330	21.11½

3d sire of

Robinette 7114	14.01

4th sire of

*Pedro Alphea 13889	15.05

SONNAMBULA 3750.
Son of Omri 2916 and Silkweed 3200.
1st sire of

Hillside Gem 16,640	20.00

SON OF ALPHEA 562.
Son of Dolphin 2d 468 and Alphea 171.
1st sire of

Pride of Corisande 5323 -	16.00
Gray Therese 5322	16.00
Silversides 3857	14.03
Silver Belle 4313	14.00

SON OF HEBE 872.
Son of Cliff 176 and Hebe 1177.
2d sire of

Gossip 6165	16.07

SON OF ROSA 663.
Son of Maxse 400 and Rosa 122.
1st sire of

Sultana 2d 11,998	- 15.04

2d sire of

Goldthread 4945	16.09

SOUTHAMPTON 117.
1st sire of

Witch Hazel 1360	14.00

2d sire of

Dot of Bear Lake 6170	19.04
Conover's Beauty 12,650	18.00
Countess Lowndes 26,874	17.08
Ida of Bear Lake 6169	16.00
Glendelia 10,524 -	- 15.12
Witch Hazel 4th 6131 -	15.05½
Mary of Bear Lake 6171	14.07

3d sire of

Troth 6139 -	16.05
Glendelia 10,524 -	15.12
Etiquette 4300 -	15.08
Honey Drop 10,033 -	14.00½
St. Nick's Flora 16,195	14.00
Bellita 4553	14.00
Elite 4299	14.00

4th sire of

Attractive Maid 16,925	16.13
Lily of Maple Grove 5079	- 16.03
Euphonia 6783	16.00½
Alfritha 6783	15.03

SOUTHEY 517.
Son of Southampton 117 and Edna 2d 809.
2d sire of

Honey Drop 10,033	14.00½

SPARKS 356.
Son of Comus 54 and Henrietta 465.
1st sire of

Haddie 921 -	16.00

SPLENDEUS 16.
3d sire of

*Sutliff's Rosy 4104	14.00

4th sire of

Hepsy 2d 12,008	17.08
Palestine 3d 1104	16.04
Cowslip 5th 849 - -	15.04
Princess of Mansfield 8070	15.02
New London Gipsey 11,667	14.08
*Churchill's Betsey 4105	14.00

SPLENDID 2.
1st sire of

Rose 3d 913	16.00

2d sire of

Maggie Mitchell (unreg.)	18.02
Lady Ives 1708 - -	18.00
Palestine 3d 1104	16.04
Copper 1979	15.07

3d sire of

Lady Ives 1708 -	18.00
Pansy of Bellewood 2d 890	18.00
Maggie 3d 3221	17.08
Dusky 2525 -	16.10
Couch's Lily 3237	16.09
Canto 7194 - -	15.12
Palestine 4644	15.08
Olie 4133 -	15.00
Rene Ogden 1568	- 15.00
Abbie Z. 14,002	14.11

Lady Ives 3d 6740 -	14.08
Lady Fanning 11,169	14.06
Fandango 12,908 -	14.03
*Belle of Bloomfield 4331	14.00
Zina 1434 - -	14.00
Pretty 2526	14.00

4th sire of

Landseer's Fancy 2876	22.07½
Kitty Potter 9893	18.05
Mirth's Blanche 19,572	17.13½
Medrena 3939 -	17.12
Maggie 3d 3221	17.08
Rosa Miller 4333 -	17.07
Maggie Rex 28,263	17.00½
Abbie Z. 3d 14,742	17.00
Attractive Maid 16,925	16.13
Dimple 3248	16.11
Pride of Corisande 5323	16.00
Cascadilla 3103 -	15.12
Myrtle 2d 211	15.12
Gold Lace 10,726 .	14.13
Louvie 3d 6159 -	14.13
Duchess of Argyle 3758	.14.13
Chloe Beach 3931 -	14.08
Lady Ives 3d 6740	14.08
Aroma 8518 -	14.07
Kate Daisy 8204 -	14.04
*Churchill's Betsey 4105	14.00
*Belle of Bloomfield 4331	14.00
Lucy Gaines' Buttercup 5058	14.00

SPLENDID 3D 125.
Son of Splendid 2 and Topsy 238.
3d sire of

Pride of Corisande 5323	16.00

SPRINGVALE 89.
Son of May Duke 90 and Mayflower 146.
3d sire of

Florinanna 9862 - -	17.05
Vieva 3d 7642	16.05

4th sire of

Florinanna 9862 -	17.05

SPRINGVALE 2D 101.
Son of Springvale 89 and Victoria 104.
3d sire of

Florinanna 9862 -	17.05

4th sire of

Mary M. Allison 6308 -	20.14
*Miskwa 15,472	19.06
Florinanna 9862 -	17.05
*Dora O. 11,703	17.03
Vieva 3d 7642	16.05

STALWART 265.
4th sire of

Nelida 2d 8227	15.02½

STANDARD 553.
1st sire of

Maid of Amboy 2929	16.01

3d sire of

Roll of Honor 13,610	14.12

STANSBURY 367.
Son of Clive 319 and Plenty 950.
2d sire of

Buttery 3502 -	- 14.01

3d sire of

Queen of De Soto 12,318	14.13

STAR F. 8364.
Son of Byron 279 and Sylvia 687.
1st sire of

Roland's Bonnie 2d 19,054	19.02

STATESMAN 2407.
Son of Cæsarea 214 J. H. B. and Sylvie Simon 4534.
1st sire of

Royal Princess 2d 12,346	17.12

ST. CLEMENT 10.
2d sire of

Cowslip 5th 849	15.04

3d sire of

Blanche 594	16.00

4th sire of

Value 2d 6844	25.02½½
Ianthe 4562 - -	19.11
White Clover Leaf 45 : 2	17.15
Rosa Miller 4333	17.07
Vixen 7591 -	17.06
Oktibbeha Duchess 4422	17.04
*Pulsatilla 7551	16.03
Nellie 1507	14.02
Lady Brown 433	14.00

STELLA BULL 189.
Son of Comet 86 and Stella 437.
4th sire of

Rose of Hillside 3866	14.03½

ST. HELIER 45.
1st sire of

Chroma 4572 -	20.06
Meine's 3d 7741	20.01
Ianthe 4562 -	19.11
Ebon Edith 10,658 -	19.01
*Cinderella St. Helier 27,241	18.09
Pyrola 4566 -	18.06
Safrano 4568 -	17.08
Auria 4567	16.13
Lutea 4563	16.03
Chenie 4570	16.00
Kalmia 4561 -	15.08
Oxali's 2d 15,631 - -	15.00
Del of Willow Farm 22,464	14.08
Pavon 12,485	14.08
Erith 4564	14.00
Helve 4565	14.00
Jule 3640	14.00
Silene 4307	14.00

2d sire of

Chroma 4572	20.06
Pyrola 4566	18.06
Volie 19,465	18.01

Safrano 4568	17.08
Kaoli 18,980 -	17.08
Reckless 3569 -	17.08
Mhoon Lady 6560	17.08
Auria 4567	16.13
Perces Lee 5538	16.10
Zithey 9184	16.07
Lesbie 9179 -	16.03
Nipheta 9180	16.00
Avis E. 9714 -	15.14
Silenta 17,685 -	15.10
Cenie Wallace 2d 6557 -	15.04½
Bessie Bradford 2d 7271	15.02
Florry Keep 6556 -	14.14
Renini 9181 - -	14.10½
Mountain Lass 12,921	14.09
Renown 13,729 -	14.06
Jeannie Platt 6005	14.04
Nannie Fitch 9143	14.04
Bintana 9837	14.03½
Flamant 11,270	14.02
Taglioni 9182	14.01
Erith 4564	14.00

3d sire of

Volie 19,465	18.01
Kaoli 18,980 -	17.08
Zithey 9184 -	16.07
Willis 2d 4461	16.03
Lesbie 9179	16.03
Nipheta 9180	16.00
Silenta 17,685 - - -	15.10
Bessie Bradford 2d 7271	15.02
Renini 9181 -	14.10½
Trenie 17,770	14.10
Renown 13,729 -	14.06
Queen of Chenango 17,771	14.06
Nannie Fitch 9143 -	14.04
Lebanon Daughter 6106	14.04
Bintana 9837 -	14.03½
Flamant 11,270 -	14.02
Lebanon Lass 6108	14.02
Taglioni 9182	14.01

4th sire of

Volie 19,465	18.01
Kaoli 18,980	17.08
Zithey 9184	16.07
Lesbie 9179	16.03
Nipheta 9180 -	16.00
Silenta 17,685	15.10
Reality 16,537	15.03½
Renini 9181	14.10½
Trenie 17,770 -	14.10
Queen of Chenango 17,771	14.06
Bintana 9837 -	14.03½
Flamant 11,270 -	14.02
Taglioni 9182 -	14.01

ST. MALO 486.
2d sire of

My Queen 4886	17.09
Bryant 4193	14.08
Myth 2837	14.06

3d sire of

Lydia Darrach 4903	17.04
La Pera 2d 13,404	14.08

ST. MALO, JR., 733.
Son of St. Malo 486 and Bright Eye 1830.
2d sire of

Lydia Darrach 4903	17.14

ST. MARTIN 1482.
Son of Stockwell 116 J. H. B. and Beauty 5311.
1st sire of

Enigma 5360	15.06

2d sire of

Mitten 13,368	15.11

3d sire of

Almah of Oakland 11,102 -	16.14
Mitten 13,368 -	15.11
Belle Thorne 13,369 -	14.11

ST. NICK 7224.
Son of Mercury 432 and Azile 1256.
1st sire of

St. Nick's Flora 16,195	14.00

STOCKWELL 116 J. H. B.
2d sire of

Gazelle 3d 6027	16.03
Enigma 5360	- 15.06

3d sire of

Mitten 13,368	15.11

4th sire of

Almah of Oakland 11,102	16.14
Mitten 13,368 -	- 15.11
Belle Thorne 13,369	14.11

STOCKWELL 396.
3d sire of

Lady Adams 2d 6529	- 15.03

STOCKWELL 2D 24 J. H. B.
Son of Noble 104 J. H. B. and Socique 68 ▲ J. H. B.
2d sire of

Reception 8557 -	19.08

3d sire of

Prize Clementine 10,322	15.12
Forget-me-not 5809 -	15.08
Deerfoot Girl 15,329	15.08
Energy 22,016	14.05

4th sire of

*Punchinello 11,875	17.11¼
Lydia of Libby 11,698	15.08
Forsaken 7520 -	15.01
Sweetrock 18,256	14.11¼
Jazel's Maid 11,011	14.06

STOKE POGIS 1259.
Son of Young Rioter 751 J. H. B. and " Essay."
1st sire of

Matilda 4th 12,816 -	16.13
La Petite Mere 2d 12,810	15.11
Marjoram 2d 12,805 -	15.00

2d sire of

Mary Anne of St. L. 9770 -	36.12½
Ida of St. Lambert 24,990	30.02½
Mermaid of St. L. 9771	25.13½
*Allie of St. L. 24,991 -	24.00
Naiad of St. L. 12,965	22.02¼

Niobe of St. L. 12,969 -	21.09½
Honeymoon of St. L. 11,221	20.05½
Rioter Pink of Berlin 23,665	19.14
Crocus of St. L. 8351	17.12
Cowslip of St. L 8349	17.12
La Belle Petite 5472 -	17.08
Brenda of Elmhurst 10,762	17.04½
Minette of St. L. 9774	17.04
*Dido Miss 8759 -	17.01
Diana of St. L. 6636	16.08
Maggie of St. L. 9776	16.03
Moth of St. L. 9775 - .	16.02
Minnie of Oxford 12,806	16.00
La Petite Mere 2d 12,810 -	15.11
Mavourneen of St. L. 9777	15.07
May Day Stoke Pogis 28,353	15.03
Jessie Brown of Maxwell 7266 -	14.07
Nora of St. L. 12,962	14.07
Cupid of Lee Farm 5997	14.06
Nancy of St. L. 12,964	14.05

3d sire of

Rioter's Nora 21,778	15.09
Carrie Pogis 22,568	15.09
Maggie Sheldon 23,583	15.03
Rioter's Ruth 14,882	14.12
Rioter's Beauty 14,894	14.00

4th sire of

Rose of St. L. 20,426	21.03½
Rioter's Beauty 14,894	14.00

STOKE POGIS 3d 2238.
Son of Stoke Pogis 1259 and Marjoram 3239.

1st sire of

Mary Anne of St. L. 9770	36.12½
Ida of St. L. 24,990	30.02½
Mermaid of St. L. 9771 -	25.13½
*Allie of St. L. 24,991	24.00
Naiad of St. L. 12,965	22.02¾
Niobe of St. L. 12,969 -	21.09½
Honeymoon of St. L. 11,221	20.05½
Rioter Pink of Berlin 23,665	19.14
Crocus of St. L. 8351	17.12
Cowslip of St. L. 8349	17.12
La Belle Petite 5472 -	17.08
Brenda of Elmhurst 10,762	17.04½
Minette of St. L. 9774	17.04
*Dido Miss 8759 -	17.01
Diana of St. L. 6636	16.08
Maggie of St. L. 9776	16.03
Moth of St. L. 9775	16.02
Minnie of Oxford 12,806	16.00
Mavourneen of St. L. 9777 - .	15.07
May Day Stoke Pogis 28,353	15.03
Nora of St. L. 12,962 -	14.07
Jessie Brown of Maxwell 7266 -	14.07
Cupid of Lee Farm 5997 -	14.06
Nancy of St. L. 12,964 -	14.05

2d sire of

Rose of St. L. 20,426	21.03½
Rioter's Nora 21,778	15.09
Carrie Pogis 22,568 -	15.09
Maggie Sheldon 23,583 -	15.03
Rioter's Ruth 14,882	14.12
Rioter's Beauty 14,894	14.00

3d sire of

Rioter's Beauty 14,894	14.00

STONEWALL JACKSON 24.
Son of Saturn 94 and Stonewall Jackson's Dam 18.

4th sire of

Favorite's Rajah Rex 16,158	15.00

ST. VALENTINE 2251.
Son of Mercury 432 and Clotho 2666.

1st sire of

Colie 8309	18.04

2d sire of

Smoky 13,733	14.09

SUCCESS 2097.
Son of Lord Ogden 69 and Belinda 1128.

1st sire of

Lady Ives 3d 6740	14.08

2d sire of

*Myra Overall 10,317	14.00

SUFFOLK 607.
Son of Comet 130 and Jersey Belle 1526.

3d sire of

Siloam 17,623	18.10
Silvia Baker 7893 -	16.04
Countess Coomasie 19,339	15.08½
Irene of Short Hills 5137	14.06½

SULTAN 58 J. H. B.
Son of Prince of Wales and Flower 58 J. H. B.

1st sire of

Lisette 483 J. H. B.	16.04
Fille de l'Air 2474	14.00

2d sire of

Garenne 1575 J. H. B.	17.08
Regina 2d 2475	14.08
Esperanza	14 03
Negress 7651	14.00

3d sire of

Chrome Skin 7881 -	20.10
Queen of Delaware 17,029	18.13
Butter Star 7799	18.04½
Panatella 4778 -	18.03
Regina 4th 12,732 - -	17.13½
Garenne 1575 J. H. B.	17.08
Faultless 12,018 -	17.05½
Faith of Oaklands 19,696	17.04
Fear Not 6059 -	17.03
Cream of Sidney 17,028	17.02½
Lucilla Kent 8892	15.10
Enigma 5360 - -	15.06
Maid of Five Oaks 7178	15.04
Fan of Grouville 7458	15.00
Walkyrie 5708	14.01

4th sire of

Princess 2d 8046	46.12½
Sultane 2d 11,373	23.08
Ona 7840 - -	20.13
Daisy of St. Peters 18,175	20.05½
Dora Neptune 20,318	20.00½
Oaklands Cora 18,853	19.09½
Reception 8557 -	19.08
Queen of Delaware 17,029	18.13
Merry Duchess 13,693	18.09½
Panatella 4778 -	18.08
Calendine 9415 -	17.09
Faith of Oaklands 19,696	17.04

Fear Not 6059 -	17.03
St. Jeannaise 15,789	16.04
Fear Not 2d 6061 -	16.02
Brunette Le Gros 9755	15.15
Fillpail 16,530	15.11
Mitten 13,368 -	15.11
Lucilla Kent 8892 -	15.10
Fan of Grouville 7458	15.00
Jenny Le Brocq 9757	14.14
Cocotte 11,958	14.12
Cosetta 15,991	14.11
Satin Bird 16,380	14.10
Blonde 2d 9268	14.04
Ballet Girl 18,750	14.01
Daisy Queen 9619	14.00

SUPERB 1956.
Son of Pierrot 2d 1669 and Myrtle 2d 211.
1st sire of

Belmeda 6229	18.12
Lida Mullin 9198	16.08
Aroma 8518	14.07
Lizzie D. 10,408	14.00

SWEEPSTAKES 682.
Son of Tam O'Shanter 381 and Susie 959.
2d sire of

Nordheim Creamer 9758	14.00

SWEEPSTAKES DUKE 1905 (76 J. H. B.).
Son of Merry Boy 61 J. H. B. and Superb
353 J. H. B.
1st sire of

Prize Clementine 10.322	15.12
Forget-me-not 5809	15.08
Deerfoot Girl 15,329	15.08
Energy 22,016 -	14.05
Florence Billot 7849 -	14.05

2d sire of

Handsome Myra 14,244	20.08
*Punchinello 11,875	17.11½
Lady Velveteen 15,771 - ⎯ -	17.02
Les Cateaux 2d 15,538	16.01
Lydia of Libby 11,698	15.03
Forsaken 7520	15.01
Sweetrock 18,256 -	14.11½
Lady Vertumnus 13,217	14.10
Jazel's Maid 11,011	14.06
La Rouge 12,405	14.05

3d sire of

St. Jeannaise 15,789	16.04
Sweet Sixteen 10,682	14.15

SYDNEY 3262.
Son of Morse 847 and Ironette 3136.
1st sire of

Florinanna 9862	17.05
Vieva 3d 7642	16.05

SYOSSET 330.
Son of Express 328 and Carrie 870.
2d sire of

Panatilla 4778	18.03

TAINTOR 70.
Son of Splendeus 16, and dam imp., by
Taintor.
3d sire of

Maggie Mitchell (unreg.)	18.02
Palestine 3d 1104 -	16.04
Cowslip 5th 849	15.04

4th sire of

Value 2d 6844 - -	25.02½
Miss Blanche 2515 (10 days)	20.09
Lady Fanuing 11,169	14.06

TAINTOR'S BULL OF 1858 306.
3d sire of

Edith 4th 817 -	14.00
Allen's Fawnette 3722	14.00

4th sire of

*Katie Kohlman 7270	23.10
Jo 5th 280	17.08
Kitty Lake 8250	15.08½
Trudie 2d 4084	15.00
Hebe 3d 3613 -	15.00
Bessie Bradford 7269	14.02
Allen's Fawnette 3722	14.00
Rioter 2d's Venus 3658 -	14.00

TAINTOR'S BULL OF 1856 307.
3d sire of

Belle of Middlefield 1516	16.03

4th sire of

Princess Bellworth 6801	15.10½

TALLY-HO 880.
2d sire of

Dark Cloud 9364	15.03½

TAMERLANE 4287.
Son of Oxoli 1922 and Ianthe 4562.
1st sire of

Trenie 17,770 -	14.10
Queen of Chenango 17,771	14.06

TAM O'SHANTER 381.
3d sire of

Beeswax 9807 -	17.05
Nordheim Creamer 9758	14.00

TANCRED 501.
Sire on I. of Jersey, dam Velvet 294.
2d sire of

Bessie S. 5002	16.00
Chenda 4599	15.09½

3d sire of

Belle of Vermillion 8798	15.02
Lillian Mostar 10,364	14.03
Sasco Belle 13,601	14.00

4th sire of

Nannie Fitch 9143	14.04

TARQUIN 750.
Sire on I. of Jersey, dam Lottie Warren 1667.
2d sire of
Lady Adams 2d 6529	15.03
Bohemian Gipsey 17,452	14.11

THE HUB 1009.
Son of Motley 515 and Bessie 139.
1st sire of
Mink 2d 3890 -	19.11
Oktibbeha Duchess 4422	17.04
Mink 3d 4868	14.09
Adora 18,569 - -	14.03

2d sire of
Mhoon Lady 6560	17.03
Julia Evelyn 6007	15.15½
Valerie 6044 -	15.13
Dairy C. 12,227	15.01

3d sire of
Marpetra 10,284	14.06
Therese M. 8364 -	14.02

THE MARQUIS 3805.
Son of Joseph 3419 and Dark Eyes 8330.
1st sire of
Grinnelle Lass 11,859	16.10

THE SQUIRE 1298.
Son of Mr. Toodles 377 and Mattie 994.
2d sire of
Jenny Dodo H. 14,448	21.08
Nibbette 11,625 -	14.07

THORNDALE 2582.
Son of Balsora 2357 and Katinka 5264.
1st sire of
Almah of Oakland 11,102 -	16.14

THOROUGH-BASS 564.
Son of Tancred 50 and Emblem 90.
2d sire of
Corn 10,504	16.02

TIM 112 J. H. B.
1st sire of
Lucy 4577	16.08

2d sire of
Souvenir 2743 J. H. B.	16.08
Denise 8281	14.04½

TIMBUCTOO 3659.
Son of Prince of Rose Croix 1893 and Quaker Girl 4551.
2d sire of
*Wabash Girl 14,550	16.00

TIP 366.
Son of Pioneer 368 and Button 953.
3d sire of
Azelda 2d 7022 :	15.02

TISQUANTUM 262 J. H. B.
1st sire of
Nervine 25,932	14.01⅛

TOCSIN 1912.
Son of "Welcome" on I. of Jersey and Twilight of South East 4547.
2d sire of
Prince's Bloom 9729	14.03

3d sire of
Grinnelle Lass 11,859	16.10

TOLEDO BOY 4641.
Son of Jo Bradley 4640 and Village Girl 5744.
1st sire of
*Wabash Girl 14,550	16.00

TOM 77 J. H. B.
Son of Welcome 172 J. H. B. and Belle 302 J. H. B.
1st sire of
Brunette Le Gros 9755	15.15

2d sire of
Floribundus 2d 14,949	18.08
Dairy Pride 4th 551 J. H. B.	16.00
Beauty 17,414 -	15.00
Saut Faluct's Rose 4805 J. H. B.	14.07

3d sire of
Belle Grinnelle 3d 16,503	14.02

TOM BROWN 3d 2940.
Son of Ike Felch 1292 and Flora F. 5544.
2d sire of
Wakena 19,721	16.00

TOM BROWN 2D 5639.
Son of Tom Brown 2940 and Notre Dame 8482.
1st sire of
Wakena 19,721	16.00

TOM DASHER 420.
Son of Albert 44 and Flora 420.
1st sire of
Jersey Cream 3151	17.00
Olie 4133 -	15.00
Creamer 2467	14.01

2d sire of
Peggy Leah 3097	18.12
May Blossom 6557 -	18.11
Duchess of Argyle 3758	14.13
Jersey Cream 2d 8519	14.12

3d sire of
Value 2d 6844 -	25.02½½
Katie Bashford 15,982	17.00
Polly Clover 7052 -	16.15
Olie's Lady Teazle 12,307	16.05
*Wabash Girl 14,550	16.00
Kitty Lake 8250	15.08½
Orphean 4636	15.07
Cowle's Nonesuch 6199 -	14.12
Lady Gray of Hill Top 2d 14,641	14.12

Gem of Sassafras 8434 -	14.03½
Lady Gray of Hill Top 3d 14,642	14.02
Hurrah Pansy 12,153	14.01½

4th sire of

Hillside Gem 16,640	20.00
Belmeda 6229	18.12
Cordelia Baker 8814	17.09
*Wabash Girl 14,550	16.00
Phyllis of Hillcrest 9067	14.12
Roll of Honor 13,610	14.12

TOMMY GREY 1099.
Son of Mercury 432 and Edith 3d 806.
2d sire of

Corn 10,504	16.02

TOM TITTLER 87.
Sire on I. of Jersey, dam Bessie 139.
3d sire of

Snowdrop F. W. 16948	14.08

TOP SAWYER 1404.
Son of Marius 760 and Emblem 90.
1st sire of

Vixen 7591	17.06
Beeswax 9807	17.05
Busy Bee 6336	16.04
Matindy 6670	16.08
Dora Doon 12,909 -	15.00
*Woodland Margaret 6215	14.10½
Opaline 7590 -	14.10
Denise 8281 -	14.04½
Fandango 12,908	14.03
Litza 6338 - -	14.03
Romp Ogden 3d 5458	14.01

2d sire of

Romping Lass 11,021	15.00
Jaquenetta 10,958 -	14.06
Variella of Linwood 10,954	14.01

TORMENTOR 3583.
Son of Khedive 108 J. H. B. and Angela 1607 J. H. B.
1st sire of

Little Torment 15,581	23.02½
Daisy Brown 12,213	17.06
Odelle Sales 15,564	16.08
Rose of Oxford 13,469	15.14½
Romping Lass 11,021	15.00
Ada Minka 15,562	14.00

TOUCHSTONE 315.
Son of Hartford 52 and Topaz 75.
2d sire of

Lady Josephine 11,560 (8 days)	19.02

3d sire of

Forsaken 7520	15.01
Coronilla 8367 -	14.09½

TROCADERO 1422.
Son of Hillhurst 1210 and Metella 3196.
1st sire of

Pansy of Bellewood 2d 8904	18.00

TROUBADOUR 481.
Son of Second Iron Duke 202 and Cybele 136.
1st sire of

Cerita of Meadowbrook 5056	17.08

TRUST 162 J. H. B.
2d sire of

Butter Star 7799 -	18.04½
Maid of Five Oaks 7178	15.04

TRUSTY 1101.
Son of Mercury 432 and Trudie 277.
1st sire of

*Katie Kohlman 7270	23.10
Lass Edith 6290	17.00
Myra 2d 6289	16.00

2d sire of

Renown 13,729	14.06

3d sire of

Reality 16,537 -	15.03½

TUBAL CAIN 1345.
Son of Castor 686 and Lady Carrie 1460.
1st sire of

Spring Leaf 5796	14.00

TYCOON 917
Son of Potomac 153 and Pauline 2130.
2d sire of

Jessie Lee of Labyrinth 5290	14.07

TYCOON, JR., 1212.
Son of Tycoon 917 and Connie 1750.
1st sire of

Jessie Lee of Labyrinth 5290	14.07

TYPHOON 77.
2d sire of

Molly 3554	16.00

3d sire of

Moss Rose of Willow Farm 5194	18.08½
Helen 3556	15.09

4th sire of

Ebon Edith 10,653 - -	19.01
Moss Rose of Willow Farm 5194	18.08½
Pavon 12,485 -	14.08

UNCAS 628.
Son of Adonis 39 and Floss 1560.
2d sire of

Rosabel Hudson 5704	15.12

UNCLE PETE 187.
4th sire of

Etiquette 4300 - -	15.06
Elite 4299 -	14.00

UNCLE PETE NO. 2 186.
Son of Uncle Pete 187 and Jessie 438.
3d sire of

Etiquette 4300	15.08
Elite 4299 -	14.00

4th sire of

Rose of Hillside 3866	14.03½

URSEL 1765.
Son of Brookside 1104 and Hebe 5th 1181.
2d sire of

Yellow Locust 10,679	14.10½

VANGUARD 845.
Son of Springvale 2d 357 and Rose 150.
2d sire of

Florinanna 9862	17.05

3d sire of

Florinanna 9862 -	17.05
Vieva 3d 7642	16.05

VERMONT 893.
Son of Governor 890 and Victorine 2238.
1st sire of

Empress 6th 3203 -	17.09¾
Goddess of Staatsburg 5252	14.08

2d sire of

Pride of Bovina 8050 -	16.09
Dorothy of Bovina 9373	15.04
Lady Adams 2d 6529	15.03

3d sire of

Almah of Oakland 11,102	16.14

VERNON 1071.
Son of Marius 760 and Velvet 294.
2d sire of

Attractive Maid 16,925	16.13

VERTUMNUS 161 J. H. B.
Son of Duke 76 J. H. B. and Coomassie 11,874.
1st sire of

*Punchinello 11,875 -	17.11½
Lady Velveteen 15,771	17.02
Les Cateaux 2d 15.538	16.01
Lady Kingscote 26,085 -	15.10
Lady Vertumnus 13,217	14.10 .
La Rouge 12,405	14.02
Lady Young 16,668 -	14.00

2d sire of

St. Jeannaise 15,789	16.04

VESPER'S ROYAL SON 2946.
Son of Iron Bank 1120 and Vesper 1395.
1st sire of

Royal Sister 12,451	14.11

VESPUCIUS 758.
Sire on I. of Jersey, dam Vesper Lass 1784.
2d sire of

Azelda 2d 7022	15.02

Cottage Lass 5332	14.08
Vespucia 17,455 - - -	14.04
Nellie Gray of Clermont 10,905	14.01

3d sire of

Vespucia 17,455 -	14.04

VICTOR 148 J. H. B.
Son of Tom 77 J. H. B. and La Petite Janne 1065 J. H. B.
1st sire of

Floribundus 2d 14,949 -	18.08
Dairy Pride 4th 521 J. H. B.	16.00
Dairy Pride 6th 21,681	16.00
Miss Huelin 22,296	14.09

VICTOR 197.
1st sire of

Pet of Maplewood Farm 4854	15.02

2d sire of

Mollie Garfield 12,172	18.07

VICTOR 3550.
Son of Pilot 3549 and Minnie 7826.
1st sire of

Jersey Belle of Scituate 7828	25.03

2d sire of

Jersey Belle of Scituate 7828	25.03
Belle of Scituate 7977	16.00
Lass of Scituate 9555	15.14
Minnie of Scituate 17,829	14.04½

3d sire of

Belle of Scituate 7977	16.00
Lass of Scituate 9555 -	15.14
Minnie of Scituate 17,829	14.04½

4th sire of

Minnie of Scituate 17,829	14.04½

VICTOR HUGO 197.
2d sire of

Sweet Briar of St. L. 5481	22.12
Melia Ann 5444 -	18.00½
Jolie of St. L. 5126 -	15.13½
Duchess of St. L. 5111 -	15.11
Nancy of St. L. 12,964	14.05
Clematis of St. L. 5478	14.03

3d sire of

Ida of St. L. 24,990	30.02½
*Allie of St. L. 24,991 -	24.00
Honeymoon of St. L. 11,221	20.05½
Variella 6337 .	18.03¾
Melia Ann 5444 -	18.00½
Cowslip of St. L. 8349 -	17.12
Brenda of Elmhurst 10,762	17 04½
Minette of St. L. 9774	17.04
Chamomilla 7552 -	16.10
Diana of St. L. 6636	16.08
Pulsatilla 7551	16.03
Carrie Pogis 22,568	15.09
Maggie Sheldon 23,583 -	15.03
May Day Stoke Pogis 28,353 -	15.03
Coquette of Glen Rouge 17,559	15.01½
Honeysuckle of St. Annes 18,674	14.14
Bonnie 2d 5742 -	14.11½
Uinta 5743	14.10
Jessie Brown of Maxwell 7266	14.07
Nora of St. Lambert 12962	14.07

Pearl of St. Lambert 5527 -   14.02
Moss Rose of St. L. 5114   14.00½
Nordheim Creamer 9758   14.00

4th sire of

Mermaid of St. L. 9771   25.13½
Naiad of St. L. 12,965   22.02½
Niobe of St. L. 12,969   21.09½
Rose of St. L. 20,426 -   21.03½
Rioter Pink of Berlin 23,665   19.14
Crocus of St. Lambert 8351   17.12
Judith Coleman 11,391 -   17.05
*Dido Miss 8759   17.01
Moth of St. L. 9775   16.02
Carrie Pogis 22,568   15.09
Maggie Sheldon 23,583   15.03
Aleph Judea 11,389 -   15.01¾
Coquette of Glen Rouge 17,559   15.01½
Honeysuckle of St. Annes 18,674 -   14.14
Rioter's Nora 14,882   14.12
Rioter's Beauty 14,894   14.00

VIGOR 1034.
Son of Mogul 532 and Cybele 136.

2d sire of

Glendelia 10,524   15.12

VOLUNTEER 1253.
Son of Quaker 887 and Victorine 2233.

2d sire of

Siloam 17,623   18.10

WABASSO 1809.
Son of Rob 874 and Princess Martie 2204.

2d sire of

Alphea Star 16,532   14.04½

WALLACE BARNES 1264.
Son of Living Storm 173 and Angeline 3247.

1st sire of

Dimple 3248   16.11

3d sire of

Hazalena's Butterfly 10,123   14.00

WANDERER 3014.
Son of Signal 1170 and Cosette 3874.

1st sire of

Fadette of Verna 3d 11,122   22.08½
Fairy of Verna 2d 10.973   20.03¾
Hilda A. 2d 11,120 -   20.00

WARWICK 264.

2d sire of

Stanstead Belle 4709   14.11½

WATERBURY 1155.
Son of Prince John 22 and Clover 2d 2902.

3d sire of

Hepsy 2d 12,008 -   17.08
Jessie Leavenworth 8248   14.00
*Sutliff's Rosy 4104 -   14.00

4th sire of

Kitty Potter 9893   18.05
Bell Rex 11,700   · _ 14.10
Webster's Pet 4103 -   14.02
*Churchill's Betsey 4105   14.00

WATERPOWER 756.
Sire on I. of Jersey, dam Purity 1408.

2d sire of

Merlette 4988   16.00

WELCOME 166 J. H. B.
Son of Noble 104 J. H. B. and Daisy 673 J. H. B.

1st sire of

Garenne 1575 J. H. B.   17.08

2d sire of

Faith of Oaklands 19,696   1047.
Fear Not 6059 -   17.03
Lucilla Kent 8892 -   15.10
Fan of Grouville 7458   15.00

3d sire of

Princess 2d 8046   46.12½
Ona 7840 - -   20.13
Daisy of St. Peters 18,175   20.05½
Dora Neptune 20,318   20.00½
Oaklands Cora 18,853   19.09½
Lactine 10,680 - ' -   17.01½
St. Jeannaise 15,789   16.04
Fear Not 2d 6061   16.02
Brunette Le Gros 9755   15.15
Lucilla Kent 8892   15.10
Jenny Le Brocq 9757   14.14
Cocotte 11,958 -   14.12
Satin Bird, 16,380   14.10
Blonde 2d 9268 -   14.04
Ballet Girl 18,750   14.01
Daisy Queen 9619   14.00

4th sire of

Oxford Kate 13,646   39.12
Little Torment 15,581   23.02½
Floribundus 2d 14,949   18.08
Daisy Brown 12,213   17.06
Princess of Ashantee 13,467   16.05
St. Jeannaise 15,789 -   16.04
Odelle Sales 15,564   16.03
Desire 9654 - -   16.03
Dairy Pride 4th 521 J. H. B.   16.00
Rose of Oxford 13,469   15.14½
Romping Lass 11,021   15.00
Beauty 17,414 -   15.00
Ada Minka 15,562   14.00

WELCOME 172 J. H. B.
Son of Welcome 166 J. H. B.

2d sire of

Daisy of St. Peters 18,175 -   20.05½
Oaklands Cora 18,853   19.09½
Brunette Le Gros 9755   15.15
Jenny Le Brocq 9757   14.14
Cocotte 11,958   14.12
Satin Bird 16,380   14.10

3d sire of

Floribundus 2d 14,949 -   18.08
Dairy Pride 4th 521 J. H. B.   16.00
Beauty 17,414   15.00

4th sire of

Belle Grinnelle 3d 16,503   14.02

WELCOME 217 J. H. B.

1st sire of

Hill's Maid of Jersey 8173   18.00

WESTCHESTER 1266.
Son of Inachus 928 and Clytemnestra 2455.
3d sire of
*Pedro Alphea 13,889	15.05

WETHERSFIELD 966.
Son of Albert 44 and Grinnella 2d 1303.
1st sire of
Lady Gray of Hill Top 6850	18.12

2d sire of
Summerline 8001	18.06
Cordelia Baker 8814	17.09
Mary Clover 9998	14.15
Lady Gray of Hill Top 2d 14,641	14.12
Deborana 4718	14.08
Jennie of the Vale 9553	14.06½
Lady Gray of Hill Top 3d 14,642	14.02

3d sire of
Olie's Lady Teazle 12,307	16.05

4th sire of
Percie 14,937	16.12
*Wabash Girl 14,550	16.00

WHIP 2638.
Son of Flash 2532 and Grinnella 4th 5926.
1st sire of
Lady Cecelia 24,821	16.01

WILLIAM HUDSON 905.
Sire on I. of Jersey, dam Nancy Peel 2384
1st sire of
Clytemnestra 2d 5868	16.00

WILLIE 12 J. H. B.
2d sire of
Joan d'Arc 2163	16.13½
Alice of Salem 5053	14.08

3d sire of
Ochra 2d 11,516	16.06½
Lustre 2062	15.08½
Buttery 3502	14.01
Witch Hazel 1360	14.00

4th sire of
Eveline of Jersey 6781	18.06
Beeswax 9807	17.05
Lily of Maple Grove 5079	16.08
Matindy 6670	16.03
Dairy Pride 4th 521 J. H. B.	16.00
Dairy Pride 6th 21,681	16.00
Rose of Oxford 13,469	15.14½
Witch Hazel 4th 6131	15.05½
Atricia 6029	15.03
Alice of the Meadows 20,748	14.12
Gold Princess 8809	14.12
*Woodland Margaret 6215	14.10½
Opaline 7590	14.10
Gilda 2779	14.06
Denise 8281	14.04½
Fandango 12,908	14.03
Litza 6338	14.03
Pixie 4115	14.00

WILLIE BOY 434.
Sire on I. of Jersey, dam Lady Mary 1148.

1st sire of
Pussie 3035	19.01

2d sire of
Welma 5942	17.08
Chenda 4599	15.09½

3d sire of
*Optima 6715	23.11
Tenella 6712	22.01½
Croton Maid 5305	21.11½
Fair Starlight 1745	17.07½
Vixen 7591	17.06
Valhalla 5300	17.00
Belle of Patterson 5664	16.07
Troth 6139	16.05
Busy Bee 6336	16.04
Œnone 8614	15.14
Edwina 6713	15.13
Fanny Taylor 6714	15.12
Etiquette 4300	15.08
Signalana 7719	15.04
Aldarine 5801	15.01½
*Queen of Maple Dale Farm 7036	14.14
*Woodland Margaret 6215	14.10½
Opaline 7590	14.10
Medrie Le Brocq 8888	14.07
Marpetra 10,284	14.06
Denise 8281	14.04½
Litza 6338	14.03
Fandango 12,908	14.03
Romp Ogden 3d 5458	14.01
La Rosa 10,078	14.00
Bellita 4553	14.00
Elite 4299	14.00

4th sire of
Fadette of Verna 3d 11,122	22.08½
Fairy of Verna 2d 10,973	20.03¾
Hilda A. 2d 11,120	20.00
Gardinier's Ripple 11,693	19.12½
Tenella 2d 19,521	18.12
Attractive Maid 16,925	16.13
Gazella 3d 9355	16.03
Genevieve Sinclair 11,157	16.02
Euphonia 6783	16.00½
Rupertina 10,409	15.12½
Reception 3d 11,025	15.08½
Alfritha 13,673	15.03
Romping Lass 11,021	15.00
Rosy Dream 9808	14.13
Euphorbia 11,229	14.09½
Jaquenetta 10,958	14.06
Lady Clarendon 3d 17,578	14.05½
Signetilia 16,333	14.03½
Variella of Linwood 10,954	14.01
Sadie's Choice 7979	14.00

YANKEE 1003 (27 J. H. B.).
*Son of Paddy 97 J. H. B. and Georgette 309
J. H. B.*
1st sire of
Kitty 5th 3849	16.11
Ida 8th 5409	14.03

2d sire of
Belle Dame 2d 22,043	15.03
Carlo's Fanny 14,591	14.00

3d sire of
Queen of De Soto 12,318	14.13
Fall Leaf 8587	14.08
Adora 18,569	14.03

4th sire of

Viva Le Brocq 13,702	17.07
Armon 10,862 -	16.13½
Queen of De Soto 12,318	14.13
Belle Grinnelle 3d 16.503	14.02

YELLOW SKIN 871.
Son o' Cliff 176 and Ariadne 608.

2d sire of

Lady Louise 4339	15.00
Gilt Edge 2d 4420 -	14.00
Gilt 4th 4208	14.00

3d sire of

Gilt Edge C. 12,223	14.03½
Sasco Belle 13,601	14.00

YORK 8.
Sire on I. of Jersey, dam Pansy 8.

2d sire of

Heartsease 503 -	15.00
Belle of Vermillion 8798	15.02
Oxalis 606	15.00
Pansy 602	14.00

4th sire of

Chamomilla 7552 -	16.10
Belle of Vermillion 8798	15.02
Oxalis 2d 15,631	15.00
Adina 1942	14.04
Helve 4565	14.00

YOUNG BARON 702.

1st sire of

Pearl Armstrong 2670	12.10
Arietta 5115	15.00

2d sire of

Duenna's Duchess 5508	16.10
Sunny Lass 6033	14.07
Muezzin 3670	14.00

YOUNG BRITON 1840.
Son of Briton 919 and Jessie of Ipswich 2883.

2d sire of

Countess of Scarsdale 18,633	14.06

3d sire of

Countess of Scarsdale 18,633	14.06

YOUNG CONCORD 1406.
Son of Concord 1405 and Molly 3554.

1st sire of

Moss Rose of Willow Farm 5194	18.08½

2d sire of

Pavon 12,485	14.08

YOUNG DAVY 661.
Son of Sir Davy 84 and Georgetta 93.

2d sire of

Grace Davy 8292	14.02

3d sire of

Bessie Ridgely 8293	14.11½

YOUNG DUKE 138.
Son of Malcolm 71 and Duchess 101.

1st sire of

My Queen 12,614	15.08

YOUNG MAJOR 214.
Son of Major 75 and Brenda 789.

2d sire of

Lucy Gray 2746	15.13

3d sire of

Jenny Dodo H. 14,448 -	21.08
Miss Blanche 2515 (10 days)	20.09

4th sire of

Ebon Edith 10,653	19.01

YOUNG PILGRIM 302.

3d sire of

Hazen's Bess 7329	24.11
Bet Arlington 8970	18.11

4th sire of

Countess Queen 13,519	18.03
Roll of Honor 13,610	14.12
Litty 8017	14.00

YOUNG PRINCE 182 J. H. B.

1st sire of

*Butterfly 18,197	14.03½

YOUNG RIOTER 751 E. H. B.
Son of Rioter 746 E. H B. and Grief.

2d sire of

Matilda 4th 12,816 -	16.12
La Petite Mere 2d 12,810	15.11
Marjoram 2d 12,805	15.00

3d sire of

Mary Anne of St. L. 9770	36.12½
Ida of St L. 24,990 -	30.02½
Mermaid of St. L. 9771	25.13½
*Allie of St. L. 24,991	24.00
Naiad of St. L. 12,965	22.02½
Niobe of St. L. 12,969 -	21.09½
Honeymoon of St. L. 11,221	20.05½
Rioter Pink of Berlin 23,665	19.14
Crocus of St. L. 8351	17.12
Cowslip of St. L. 8349	17.12
La Belle Petite 5472 -	17.08
Brenda of Elmhurst 10,762	17.04½
Minette of St. L. 9774	17.04
*Dido Miss 8759 -	17.01
Diana of St. L. 6636	16.08
Maggie of St. L. 9776 -	16.03
Moth of St. L. 9775 -	16.02
Minnie of Oxford 12,806	16.00
La Petite Mere 2d 12,810	15.11
Mavourneen of St. L. 9777	15.07
May Day Stoke Pogis 28,353	15.03
Marjoram 2d 12,805 -	15.00
Jessie Brown of Maxwell 7266	14.07
Nora of St. L. 12,962 -	14.07
Cupid of Lee Farm 5997	14.06
Nancy of St. L. 12,964	14.05

4th sire of

Rioter's Nora 21,778	15.09
Carrie Pogis 22,568	15.09
Maggie Sheldon 23,583	15.03
Rioter's Ruth 14,882	14.12
Rioter's Beauty 14,894	14.00

YOUNG SIR DAVY 3034.
Son of Young Davy 661 and Susan 1658.

1st sire of

Grace Davy 8292 -	14.02

2d sire of	
Bessie Ridgely 8293	14.11½

YOUNG YANKEE 62 J. H. B.
Son of Yankee 1003 (27 J. H. B.) and Vir-
ginia 19 J. H. B.
3d sire of

*Lady Golddust 2d 19,861	14.01

ZANY 551.
Son of Lopez 313 and Pet 2d 747.
2d sire of

Orphean 4636	-	15.07

3d sire of

Belmeda 6229	.	18.12
Gem of Sassafras 8434		14.03½

ZEUS 2684.
Son of Bismarck 1428 and Oweenæ 3814.
1st sire of

Pyrrha 6100		16.14½

2d sire of

Nimble 23,335		14.10

TESTS OF 14 DAYS OR LONGER.

	Days.		Lbs.	oz.
Valhalla 5300	14		34	00
King's Trust 18,946	15		36	06½
Maggie Rex 28,823	21		47	08½
Princess 2d 8046	28		107	03
Lady Fair 1765	28	-	86	12
Effie 885 A. H. B.	30		98	00
Eurotas 2454 -	30	-	88	06
Abbie Z. 14,002 -	30		61	02
Mary Anne of St. L. 9770	31		106	12½
Bomba 10,380 -	31	-	89	14
Jeannette Montgomery 5177 -	31		89	00
Thorndale Belle 3d 10,459 -	31		89	00
Jersey Queen of Barnet 4201 A. H. B.	31		84	05
Dora Neptune 20,318	31		83	06½
Daisy Brown 12,213	31		73	04½
Couch's Lily 3237	31	-	71	00
Oak Leaf 4769 -	31		63	04
Rosebud of Belle Vue 7702	31		60	04½ -
Robema 3840 - -	31		54	00
Little Torment 15,581	56	-	83	05
Landseer's Fancy 2876	60		180	14
Lady Mel 2d 1795	61		183	00
Fair Lady 6728	62		150	04½
Bomba 10,830 -	62		174	03
Mollie Garfield 12,172	62		163	00
Eva 883 A. H. B. - -.	138		281	00
Duchess of Bloomfield 3653	8 mos. & 7 days		501	04
Flora 113 -	50 weeks	.	511	02
Masena 25,732 (Rated)	1 year		902 08	
Mary Anne of St. L. 9770 -	1 "		867 14½	
Jersey Queen of Barnet 4201 A. H. B.	1 "		851	01
Eurotas 2454	1 "		778	01
Pansy 1019 -	1 "		574	08
Webster's Pet 4108	1 "		429	00

INDEX.

NOTE.—Having mislaid some papers on which I had notes of tests, with the authorities, I am obliged to leave some spaces blank, or delay the printer too long to hunt them out. In the next edition of this work, about October, I shall try and have as many of the tests backed by affidavits as possible, and request that new tests sent me for publication then, be in that form. In reading the index, first is the name, next H. R. number, next (*in brackets*) the record, then the pages where the name will be found in the body of the book.
F. M. C.

Beauty of the Grange 7502, (23.09), Orestes Pierce, 25

Beeswax 9807, (17.05), ——, 11, 21, 22, 27, 39, 43, 67, 69, 72

Bella Delaino 10,356, (14.02), S. L. Hoover, 43, 48

Bella Donna 148 J. H. B., (16.10), J. H. Walker, 11

Belle Dame 2d 22,043, (15.03), Francis Le Brocq, 10, 27, 47, 72

Belle Grinnelle 4073, (18.08), S. W. Robbins, 2, 29, 42, 58, 67

Belle Grinnelle 3d 16,503, (14.02), E. J. Robbins, 2, 8, 30, 36, 37, 42, 58, 71, 73

Belle Hartford 2718, (15.00), Jas. A. Hayt, 58

*Belle of Bloomfield 4331, (14.00), J. H. Walker, 45, 53, 55, 64

*Belle of Inda 3867, (15.01½), ——, 11, 20, 21, 39, 53, 55, 60

Belle of Middlefield 1516, (16.03), Lyman A. Mills, 20, 33, 42, 67

*Belle of Milford 7445, (14.07), Jos. Kiplinger & Co., 4, 11, 21, 28, 44, 55, 59

Belle of Ogden Farm 1570, (14.00), Jno. H. Freeman, 6, 10, 18, 19, 23

Belle of Patterson 5664, (16.06), W. J. Chinn, 2, 24, 30, 39, 49, 62, 72

*Belle of Saybrook 6875, (15.01½), ——, 50, 58

Belle of Scituate 7977, (16.00), Chas. O. Ellms, 17, 50, 51, 70

Belle of Vermillion 8798, (15.02), J. H. Walker, 10, 17, 22, 27, 31, 36, 51, 52, 67, 73

Belle Thorne 13,369, (14.11), ——, 3, 4, 23, 31, 39, 42, 45, 65

Bellini's La Biche. 15,091, (14.14½), W. Simpson, 5, 16, 50, 51, 52, 55, 57

Bellini's Maid 15,170, (15.01½), W. Simpson, 5, 50, 51, 52, 55, 57

Bellita 4453, (14.00), J. H. Taylor, 12, 25, 39, 63, 72

Bell Rex 11,700, (14.10), Moulton Bros., 2, 6, 10, 13, 30, 36, 40, 56, 58, 60, 71

Belmeda 6229, (18.12), G. R. Dykeman, 6, 7, 28, 36, 40, 44, 50, 51, 67, 69, 74

Bergerelia 15,546, (14.01¾), T. A. Havemeyer, 58

Bertha Morgan 4770, (19.06), Edward Worth, 6, 36

Bessie Bradford 7269, (14.02), L. S. Sprague, 32, 34, 39, 41, 61, 67

Bessie Bradford 2d 7271, (15.02), Alice M. Bradford, 7, 33, 34, 39, 41, 65

Bessie Ridgely 8293, (14.11½), C. S. S. Baron, 14, 26, 32, 59, 62, 73, 74

Bessie S. 5002, (16.00), ——, 7, 22, 32, 36, 41, 61, 67

Bet Arlington 8970, (18.11), N. C. Stoughton, 20, 22, 43, 47, 49, 73

Bettie Dixon 4527, (15.00), W. B. Montgomery, 2, 11, 12, 32, 44, 55

Beulah of Baltimore 3270, (14.06½), Clark & Jones, 8, 14, 52, 53, 62

Bintana 9837, (14.03½), Joseph Gavin, 5, 45, 48, 51, 65

Birdie Le Brocq 17,263, (14.00), I. D. Risher, Sale Catalogue, 35

Black Diamond's Queen 11,865, (15.08), I. D. Risher, Sale Catalogue, (——),

Blanche 594, (16.00), O. S. Hubbell, 20, 42, 49, 64

Blonde 2d 9268, (14.04), Wm Rolph, 8, 33, 35, 44, 46, 67, 71

Bloomfield Lady 6912, (14.12), J. H. Walker, 2, 29, 31, 40, 45, 53, 55

*Blossom of Hanover 13,655, (17.08), Hiram Hitchcock, 7, 18, 32, 33, 57, 61

Blossy Reynolds 6082, (16.03½), G. H. Reynolds, 20, 21, 28, 43, 47

Bohemian Gipsey 17,452, (14.11), W. Simpson, 11, 50, 68

Bomba 10,330, (21.11½) Com. A. J. C. C., 19, 33, 41, 57, 60, 63, 74

Bonfanti 388, (14.00), C. Wellington, (——)

*Bonmari 7019, (14.00),——, 20, 39, 53, 60

Bonnie 2d 5742, (14.11½), S. E. Gillett, 4, 37, 70

Bonnie Yost 7943, (18.02), M. M. Gardner, 12, 35, 37, 42, 50, 51, 52, 53, 55, 59

Bounty 1606, (14.00), Beech Grove Farm, 10, 53, 60

Brenda 3025 J. H. B., (14.00), ——, 20, 35

* Brenda of Elmhurst 10,672, (17.04½),V. E. Fuller, 18, 36, 66, 70, 73

* Brightness 2211, (14.00), Moulton Bros., 2, 10, 29, 40, 53

* Brightness 3d 14,824, (15.05), Moulton Bros., 2, 10, 29, 30, 40

Brunette Lass 1780, (15.10), W. J. Webster, (——)

Brunette Le Gros 9755, (15.15), S. W. Robbins, 46, 67, 68, 71

Bryant 4193, (14.08), Geo. E. Bryant, 47, 65

Buckeye Lass 10,355, (14.04), S. L. Hoover, 17

Busy Bee 6336, (16.04), W. E. Oates, 21, 22, 39, 43, 69, 72

Buttercup 17,285, (16.08), ——, 60

Butterfly 18,197, (14.03½), ——, 73

Butter Star 7799, (18.04½), Campbell Brown, 9, 54, 66, 69

Buttery 3502, (14.01), Campbell Brown, 11, 12, 14, 37, 52, 64, 72

Calendine 9415, (17.09), F. C. Sayles, 5, 42, 46, 55, 56, 66

Callie Nan 7959, (16.02), Campbell Brown, 6, 9, 18, 19, 23, 31, 55

Callington 22,021, (15.10), Mrs. Susan Le Gresley, (——)

Calypris 5948, (15.04½), Geo. Jackson, 18, 19, 28, 39, 45, 55, 56

Canto 7194, (15.12), T. J. Hand, 1, 6, 7, 23, 36, 37, 44, 50, 63

Carlo's Fanny 14,591, (14.00), Mrs. D. B. Judson, 10, 27, 47, 72

Carrie 3894, (16.08), J. V. N. Willis, 63

Carrie Pogis 22,568, (15.09), V. E. Fuller, 18, 37, 48, 66, 70, 71, 73

Caroline 12,019, (14.08), J. M. Richmond, 12, 47, 48, 56

Cascadilla 3103, (15.12), ——. 9, 10, 12, 14. 21, 35, 37, 38, 40, 64

Ceccola 13,608, (16.13), W. Simpson, 10

Celia Belle 5865, (14.03), Campbell Brown, 6, 14, 16, 20, 23, 26, 30, 38, 44, 51, 59

Ceule Wallace, 2d 6557 (15.04½), W. B. Montgomery, 5, 12, 32, 51, 55, 65

Cerita of Meadow Brook, 5056, (17.08), Newton T. Beale, 20, 28, 42, 43, 51, 52, 61, 69

Chamomilla 7552. (16.10), J. T. & W. S. Shields, 4, 11, 37, 51, 52, 70, 73

Champion Chloe 12,255, (15.05½), Abram M. Turner, 6, 10, 21, 31, 35, 36, 40, 53

Charmer 4771, (14.12), Henry C. Kelsey, 21, 28, 43, 51, 56, 60, 61

Chenda 4599, (15.09½), Campbell Brown, 39, 67, 72

Desire 9654, (16.03), D. Blampied, 33, 35, 45, 46, 71

Diana of St. Lambert 6686 (16.08), W. D. Reesor, 18, 37, 57, 66, 70, 73

*Dido Miss 8759, (17.01), W. H. Corning, 8, 18, 37, 66, 71, 73

Dimple 3248, (16.11), G. W. Felter, 10, 35, 40, 53, 60, 64, 71

Dolly of Lakeside 10,824, (14.08), J. H. Walker, 15, 16, 17, 42, 43, 51

Dom Pedro's Julian 8631, (16.00), Paul Ballest, 18, 28, 32, 48

Dora Doon 12,909, (15.00), ——, 5, 12, 38, 39, 50, 52, 55, 56, 69

Dora Neptune 20,318, (20,00½), C. E. Rowley, 29, 45, 46, 66, 71, 74

*Dora O. 11.703, (17.03), A. W. Cozart, 4, 9, 14, 23, 38, 53, 56, 64

Dorothy of Bovina 9373, (15.04), W. L. Rutherford, 5, 16, 21, 24, 29, 30, 44, 50, 70

Dot of Bear Lake 6170, (16.04), Jno. C. Drake, 26, 47, 53, 54, 62, 68

Duchess Caroline 3d 6039, (15.08), W. B. Montgomery, 2, 27, 44. 55

Duchess of Argyle 3758, (14.13), E. S. Henry, 2, 13, 14, 16, 21, 29, 30, 40, 43, 64, 68

Duchess of Bloomfield 3653, (20.00½), Campbell Brown, 14, 38, 57, 58, 74

Duchess of St. Lambert 5111, (15.13), V. E. Fuller, 36, 70

Dudu of Linwood 8336, (16.15), ——, 5, 6, 7, 12, 37, 50, 51, 52, 55, 56, 59

Duenna's Duchess 5508, (16.10), G. H. & H. A. Grinnell, 19, 73

Dusky 2525, (16.10), J. B. Williams, 2, 29, 57, 63

Ebon Edith 10,653, (19.01), C. R. C. Dye, 17, 19, 38, 40, 64, 69, 73

Eclipse 14,427, (15.12), H. E. Alvord, 35

Edith 4th 817, (14.00), ——, 32, 39, 61, 67

Edwina 6713, (15.13), J. B. Wade, 2, 30, 39, 49, 62, 72

Effie 523, (19.11), J. H. Walker, 14, 39, 74

Elinor Wells 12,068, (14.00), Beech Grove Farm, 11, 20, 35, 39, 54

Elite 4299, (14.00), ——, 12, 13, 25, 33, 39, 63, 69, 70, 72

Elmora Mostar 15,955, (14.00), Jas. Cloud & Son. 11, 12, 19, 24, 33, 51, 52, 57

Elsie Brown 4021, (14.06½), J. H. Walker, 50.

Elsie Lane 13,302, (15.12), L. A. Mills, 2, 4, 7, 10, 13, 35, 56, 58

Embla 4799, (17.08), C. R. C. Dye, 9, 14, 23, 27, 45, 53, 59

Empress 6th 3203, (17.09¾), McV. Barnard, 23, 24, 44, 70

Energy 22,016, (14.05), J. H. Walker, 41, 46, 65, 67

Enid 2d 10,782, (14.07½), H. M. Howe, 6, 20, 55

Enigma 5360, (15.06), Edwin Thorne, 23, 65, 66

Epigæa 4631, (14.07), ——, 5, 21, 28, 34, 51, 59, 61

Erith 4564, (14.00), O. S. Hubbell, 5, 45, 51, 64, 65

Esperanza, (14.03), J. H. Walker, 47, 66

Estrella 2831, (14.12), J. S. Wells, 12, 17, 32, 41, 60, 61

Etiquette 4300, (15.08), Orestes Pierce, 12, 13, 25, 33, 39, 63, 69, 70, 72

Eugenie 792, (14.00), J. H. Walker, (——)

Eugenie 2d 1623, (14.00), S. C. Colt, 58

Euphonia 6783, (16.00½), A. J. C. C. Com., 24, 26, 29, 57, 61, 63, 72

Euphorbia 11,229, (14.09½), J. B. Wallace, 2, 3, 6, 16, 17, 24, 37, 39, 47, 58, 62, 72

Euphrates 9778 (16.05), T. W. Suffern, 32, 57, 61

Eureka McHenry 8341, (14.00), A. E. Kapp, 14, 26, 56

Europa 121, (18.06), I. D. Risher, Sale Catalogue (——)

Europa 176, (15.00), G. W. Harris for R. M. Hoe, 32, 61

Europa 558, (17.06), J. H. Walker, (——)

Eurotas 2454, (22.07), A. B. Darling, 32, 57, 61, 74

Eva 883 A. H. B., (14.00), ——, 39, 74

Eveline of Jersey 6781, (18.06), E. L. Clarkson, 11, 19, 24, 72

Fadette of Verna 3d 11,122, (22.08½), G. W. Farlee, 2, 24, 39, 62, 71, 72

Faith of Oaklands 19,696, (17.04), V. E. Fuller, 46, 54, 66, 71

Fair Lady 5184, (19.00), Columbia Jersey C. C., 28, 74

Fair Starlight 1745, (17.07⅓), David Strong, 9, 23, 34, 58, 72

Fairy 10, (17.00), O. S. Hubbell, (——)

Fairy of Verna 2d 10,978, (20.03¾), G. W. Farlee, 2, 24, 39, 62, 71, 72

Fall Leaf 8587, (14.08), W. E. Oates, 26, 35, 36, 50, 52, 54, 59, 72

Fancy 9, (15.06), ——, (——)

Fancy Juno 6086, (15.10), R. S. Strader, 12, 30, 52, 54, 56

Fandango 12,908, (14.03), W. Simpson, 1, 6, 12, 37, 39, 44, 64, 69, 72

Fanny Taylor 6714, (15.12), Jno. Middleton, 2, 30, 39, 49, 50, 62, 72

Fan of Grouville 7458, (15.00), Beech Grove Farm, 46, 54, 60, 66, 67, 71

Fantine 1271, (15.06), H. M. Howe, 55

Faultless 12,018, (17.05½), W. Simpson, 49, 54, 55, 66

Faustine 10,854, (14.14½), W. Simpson, 4, 18, 31, 32, 41, 43, 61

Favorite of the Elms 1656, (16.04), Wm. S. Taylor (——)

Favorite's Rajah Rex 16,153, (15.00), Jos. Kiplinger & Co., 2, 5, 13, 19, 30, 42, 55, 56, 58, 62, 66

*Fawnette of Woodstock 3710, (17.08), J. H. Walker, 4, 6, 15, 23, 30, 49

Fear Not 6059, (17.10), Woodside Farm, 8, 46, 47, 54, 60, 66, 67, 71

Fear Not 2d 6061, (16.02), Woodside Farm, 7, 8, 46, 54, 60, 67, 71

Fidelia 5817, (14.00), Chas. L. Sharpless (——)

*Filbert 3630, (15.12), F. R. Starr, 10, 27, 35, 49, 49

Fille de l'Air 2474, (14.00), J. H. Walker, 47, 54, 56

Fillpail 16,530, (15.11), W. Simpson, 18, 28, 32, 41, 49, 55, 61, 67

Flamant 11,270, (14.02), R. W. Curtis, 11, 21, 48, 65

Flora 113, (50 weeks), (511.02), ——, 74

Flora of St. Peter's 8622, (16.05), Wm. Crozier, (——)

Florence Billot 7849, (14.05), ——, 67

Floribundus 2d 14,949, (18.08), L. L. Tozier, 46, 68, 70, 71

Florinanna 9862, (17.05), ——, 43, 52, 64, 67, 70

Jeannette Montgomery 5177, (20.00), J. H. Walker, 3, 4, 45, 54, 59, 74

Jeanne Le Bas 2476, (15.08), H. Borden-Bowen, 31, 54

Jeannie Platt 6005, (14.04), Lyman A. Mills, 2, 13, 30, 45, 56, 57, 58, 65

Jennie (not reg.), (18.03), M. J. Brown, (———)

Jennie 766, (14.09), W. B. Dinsmore, (———)

Jennie of the Vale 9553, (14.06½), H. W. Douglas, 2, 7, 25, 27, 30, 33, 50, 58, 72

Jenny 287, (17.00), J. H. Walker, (———)

Jenny Dodo H. 14,448, (21.08), W. H. Blasdell, 1, 15, 17, 20, 27, 38, 43, 47, 68, 73

Jenny Le Brocq 9757 (14.14), S. W. Robbins, 27, 46, 67, 71

Jersey 3260 (15.06), Woodside Farm, 13, 17, 38, 59

Jersey Belle of Scituate 7828, (25.03), C. O. Ellms, 15, 17, 51. 70

Jersey Cream 3151, (17.00), D. B. De Wolf, 2, 29, 40, 49, 68

Jersey Cream 2d 8519, (14.12), H. G. Westlake, 2, 30, 34, 40, 58, 68

Jersey Queen of Barnet 4201 A. H. B., (19.12), A. B. Darling, 9, 17, 19, 39, 59, 74

Jessie Brown of Maxwell 7266, (14.07), S. M. Neel, 18, 37, 66, 70, 73

Jessie Leavenworth 8248, (14.00), Warren F. Daniel, 6, 10, 16, 21, 35, 36, 40, 53, 56, 60, 71

Jessie Lee of Labyrinth 5290, (14.07), Jas. Crook, 8, 14, 19, 21, 26, 52, 62, 69

Jewel 3d A. H B., (15.04), ———, 39

Joan d'Arc 2163, (16.13½), J. H. Walker, 7, 11, 42, 54, 72

Jo 5th 280, (17.08), J. W. Vance, 17, 26, 33, 53, 58, 60, 67

Jolie of St. Lambert 5126, (15.13½), W. A. Reburn, 16, 34, 36, 70

Judith Coleman 11,391, (17.05), Rutherford Douglas, 8, 30, 31, 37, 56, 71

Judy 691, (19.00), J. H. Walker, 13, 38, 43

Jule 3640, (14.00), O. H. Platt, 3, 45. 59, 64

Julia 3893, (16.08), W. A. Conover, 62

Julia Walker 10,138, (15.12), Thos. Fitch, 34, 50, 51

Julia Evelyn 6067 (15.15½), W. B. Montgomery, 2, 11, 29, 38, 43, 44, 55, 68

Kalmia 4561, (15.08), O. S. Hubbell, 4, 5, 36, 45, 51, 64

Kaoli 18,980, (17.08), S. B. Wheeler, 5, 45, 48, 51, 65

Kate Daisy 8204, (14.04), L. M. Fair, 4, 6, 7, 10, 16, 32, 40, 49, 50, 55, 58, 61, 64

Kate Gordon 8387, (15.15), M. C. Campbell, 14, 46, 49, 50, 51, 52, 54, 59

Katie Bashford 15,982, (17.00), ———, 2, 3, 6, 20, 34, 42, 50, 58, 61, 68

*Katie Kohlman 7270, (23.10), J. H. Walker, 24, 32, 33, 39, 41, 61, 67, 69

Katy Didn't 2734, (14.12), Jas. A. Hayt, (———)

King's Trust 18,946, (18.00), S. M. Burnham, 33, 74

Kitty 5th 3849, (16.11), W. B. Dinsmore, 72

Kitty Clover 1113, (14.00), Jas. A. Hayt, (———)

Kitty Colt 2213, (15.09½), W. B. Montgomery, 2, 29

*Kitty Lake 8250, (15.08½), Moulton Bros., 2, 10, 16, 29, 30, 35, 40, 58, 67, 68

Kitty Potter 9893, (18.05), C. G. Grannis, 3, 10, 40, 50, 56, 61, 64, 71

Kosi 3431, (16.02), W. H. Walrath, (———)

La Belle Petite 5472, (17.08), Cooper & Maddux, 9, 18, 57, 66, 73

Lactine 10,680, (17.01½), . M. Howe, 15, 24, 28, 31, 38, 46, 71

Lady Adams 2d 6529, (15.03), W. B. Dinsmore, 24, 43, 44, 54, 65, 68, 70

Lady Alice of Hillcrest 7450, (16.03), James Crook, 25, 31, 32, 33, 36, 42, 53, 58, 61

Lady Bidwell 10,303, (15.12), Geo. S. Phelps, 7, 9, 12, 15, 27, 37, 49

Lady Bloomfield 4704, (14.12½), Jno. B. Mills, 57

Lady Bountiful 17,946, (14.00), ———, (———),

Lady Brown 433, (14.00), S. W. Robbins, 10, 31, 40, 53, 64

Lady Brown 4th 6911, (14.12), James Woodruff, 2, 10, 31, 40, 53

Lady Bowen 354, (16.08), Jas. Cloud & Son, (———),

*Lady Caroline 3674, (14.14), F. M. Carryl, 29

Lady Cecelia 24,821, (16.01), ———, 2, 22, 42, 50, 58, 72

Lady Clarendon 3d 17,578, (14.05½), J. Horatio Earle, 2, 9, 39, 62, 72

Lady Cloud 19,358, (16.10), Jno. A. Middleton, 12, 22, 30, 54, 56

Lady Fair 1765, (14.12), W. L. Conyngham, 74

Lady Fanning 11,169, (14.06), Jos. Kiplinger & Co., 23, 50, 51, 64, 67.

*Lady Golddust 2d 19,861, (14.01), D. F. Appleton, 19, 28, 30, 32, 33, 41, 57, 60, 74.

Lady Gray of Hilltop 6850, (18.12), R. J. Fair, 2, 6, 14, 29, 30, 36, 59, 72

Lady Gray of Hilltop 2d 14,641, (14.12), L. M. Fair, 2, 6, 10, 30, 40, 59, 68, 72

Lady Gray of Hilltop 3d 14,642 (14.02), L. M. Fair, 2, 6, 10, 30, 36, 40, 59, 69, 72

Lady Hayes 10,136, (15.12), Thos. Fitch, 34, 50, 51

Lady Ives 1708,‖ (18.00), Lawson Ives, 1, 55, 63

Lady Ives 3d 6740, (14.08), J. H. Walker, 4, 6, 7, 37, 44, 55, 64, 66

Lady Jane of St. Peters 7475, (15.08), ———, (———)

Lady Josephine 11,560 (8 days), (19.02), E L. Brigg, 4, 5, 13, 14, 26, 36, 42, 48, 47, 52, 56, 60, 69

Lady Kingscote 26,085, (15.10), J. S. Shallcross, 8, 44, 70

Lady Louise 4339, (15.00), R. G. Skiff, 12, 17, 25, 32, 33, 60, 73

Lady Love 2d 2212, (16.08), Louis Brush for W. Simpson, 2, 17, 26, 29, 38, 46

Lady Mel 2d 1795, (21.00), Chas. F. Mills, 2, 10, 29, 40, 53, 74

Lady of Bellevue 7705, (15.11), M. M. Gardner, 34, 36

Lady of the Isles 2d 16652, (22.08), Francis Le Brocq (———)

Lady Oaks 2d 5246, (15.02), Jno. E. Phillips, 4

Lady Oxford 4860, (10 days), (22.02), T. C. Murphy, 22

Lady Penn 5314, (16.00), J. H. Walker, 4, 6, 16, 24, 33, 35, 42, 48, 55

Lady Velveteen 15,771, (17.02), W. R. McCready, 8, 44, 67, 70

Lady Vertumnus 13,217, (14.10), W. Simpson, 8, 44, 67, 70

Lady Young 16,668, (14.00), I. F. Johnson, 8, 44, 70

Malope 2d 11,923, (15.10), W. Simpson, 8, 11, 32, 33, 40, 41, 61

Mamie Coburn 3798, (17.08), Mrs. J. B. Ritzniger, 11, 13, 21, 39, 54, 55, 61

Maple Leaf 4768, (14.12), Jno. D. Wing, 6, 16, 22, 33, 37, 47, 61

Marie S. 12,043, (15.06), W. Simpson, 22

Maritina 12,039, (16.03½), W. Simpson, 22

Marjoram 3239, (16.00), T. S. Cooper, (———)

Marjoram 2d 12,805, (15.00), Cooper, Maddux & Co., 65, 73

Marpetra 10,284, (14.06), C. E. Douglas, 2, 3, 11, 24, 32, 39, 43, 44, 68, 72

Mary Anne of St. Lambert 9,770, (36.12½), Com. A. J. C. C., 8, 18, 37, 57, 65, 66, 73, 74

Mary Clover 9998, (14.15), C. J. Wemple, 2, 5, 10, 14, 30, 35, 36, 37, 40, 43, 58, 60, 72

Mary Jane of Bellevue 6956, (17.07), V. L. Kirkman, 12, 47, 48, 56

Mary M. Allison 6308, (20.14), C. W. H. Eicke, 21, 28, 38, 42, 51, 53, 64

Mary of Bear Lake 6171, (14.07), Chester Bordwell, 26, 47, 53, 54, 62, 63

Mary of Gilderoy 11,219, (14.04), H. M. Howe, 24, 28, 31, 38, 42

*Mary Walker 11,303, (18.12), Jos. Kiplinger, 34, 50, 51

Marvel 13,734, (15.01), W. Simpson, 4, 18, 32, 33, 41, 43, 61

Masena 25,732, (20.07), P. P. Paddock, 16, 33, 59, 60, 74

Matilda 2238, (17.00), T. S. Cooper, (———)

Matilda 4th 12,816, (16.12), T. S. Cooper, 65, 73

*Matindy 6670, (16.03), ———, 6, 11, 18, 19, 23, 28, 39, 69, 72

Maudine of Elmwood 9718, (16.15), W. B. Dugger, 12, 22, 27, 36, 43, 52

Maud Lee 2416, (23.00), F. W. Tanner, 15, 31, 59, 60

Mavourneen of St. L. 9777, (15.07), V. E. Fuller, 18, 66, 73

May Blossom 5657, (18.11), W. Simpson, 2, 27, 30, 35, 40, 68

May Day Stoke Pogis 28,353, (15.03), Geo. Smith, 18, 37, 66, 70, 73

May Fair 5184, (16.07), Columbia Jersey C. C., 25

Medrena 3939, (17.12), Beech Grove Farm, 4, 6, 18, 19, 23, 41, 43, 45, 64

Medrie Le Brocq 8888, (14.07), G. S. Williams, 18, 35, 39, 41, 43, 72

Medusa 3330 J. H. B., (15.12), J. H. Walker, (———)

Meines 3559, (14.00), J. H. Walker, 15, 17, 19, 23, 26, 39, 46

Meines 3d 7741, (20.01), V. E. Fuller, 17, 23, 39, 64

Melia Ann 5444, (18.00½), J. H. Walker, 36, 70

Merlette 4988, (16.00), J. D. Conner, 59, 71

Memento 1913, (14.05), Thomas Beer, 14, 34, 51, 53

Mermaid of St. L. 9771, (25.13½), Com. A. J. C. C., 8, 18, 34, 36, 37, 65, 66, 71, 73

Merry Burlington 7600, (15.04), J. T. & W. S. Shields, 26, 31, 41, 42, 52, 53

Merry Duchess 13,693, (18.09½), F. C. Sayles, 46, 48, 56, 66

Metah's Queen 4886, (17.09), George E. Bryant, 47

Mboon Lady 6560, (17.03), W. B. Montgomery, 5, 29, 38, 43, 45, 55, 65, 68

Miami Prize 8100, (14.00), ———, 7, 11, 14, 24, 34, 44, 52

*Mica 1983, (15.12), R. Goodman, 7, 12, 26, 27, 63

Milkmaid of Burr Oaks 9035, (14.05), T. Bacon, 12, 14, 15, 22, 28, 31, 41, 47, 50

Milky Way 18,865, (17.08½), W. Simpson, (———)

Milkweed (1673 J. H. B.) 16,402, (14.07), Edward Worth, (———)

Mineola of Elmarch 8229, (15.15), ———, 19, 50

Minette of St. Lambert 9774, (17.04), Wm. Rolph, 18, 37, 66, 70, 73

Mink 2d 3890, (19.11), W. B. Montgomery, 29, 38, 43, 68

Mink 3d 4868, (14.09), W. B. Montgomery, 29, 38, 43, 68

Minnie 2386, (15.12), H. Mead, (———)

Minnie Lee 2d 12,941, (14.03), W. B. Montgomery, 3, 6, 10, 21, 35, 36, 40, 44, 55, 58

Minnie of Oxford 12,806, (16.00), J. H. Walker, 18, 66, 73

Minnie of Scituate 17,829, (14.04½), Orestes Pierce, 16, 17, 20, 50, 51, 70

Mirtha 3437, (17.13½), Campbell Brown, 6, 14, 26, 41, 43, 52, 53

Mirth's Blanche 19,572, (17.13½), J. H. Walker, 6, 10, 14, 26, 35, 49, 52, 53, 60, 64

Mischief Le Brocq 7680, (15.00), A. E. Kapp, (———)

*Miskwa 15,472, (19.06), A. W. Cozart, 4, 9, 23, 38, 53, 58, 64

Miss Baden Baden 14,760, (14.14½), R. S. Strader, 3, 15, 16, 20, 21, 24, 27, 30, 44

Miss Blanche 2515, (10 days), (20.09), Z. C. Luse & Son, 13, 38, 44, 57, 67, 73

Miss Browny 7288, (16.13), ———, 8, 44

Miss Huelin 22,296, (14.09), T. A. Havemeyer, 70

Miss Vermont 7698, (16.05), W. R. McCready, 35

Miss Willie Jones 6918, (16.04), W. S. Taylor, 4, 16, 18, 21, 31, 33, 43

Mitten 13,368, (15.11), ———, 31, 39, 42, 45, 65, 67

Moggie Bright 25,891, (16.06), W. R. McCready, 8

Mollie Brown 7831, (16.00), Edward Worth, 6, 36

Mollie Garfield 12,172, (18.07), F. S. Peer, 5, 31, 51, 52, 57, 70, 74

Molly 3554, (16.00), ———, 15, 16, 17, 69

Monmouth Duchess 3895, (14.07), W. A. Conover, 26, 53, 62

Morlacchi 2725, (14.00), Tenn. Hosp. for Insane, 20

Moss Rose of St. Lambert 5114, (14.00½), W. Rolph, 8, 37

Moss Rose of Willow Farm 5194, (18.08½), J. H. Walker, 15, 16, 17, 26, 69, 71, 73

Moth of St. Lambert 9775, (16.02), Wm. Rolph, 8, 18, 37, 57, 66, 71, 73

Mountain Lass 12,921, (14.09), ———, 2, 5, 44, 45, 51, 55, 58, 65

Muezzin 3670, (14.00), D. A. Givens, 14, 35, 42, 51, 53, 73

Myra 2d 6289, (16.00), Thos. H. Faile, 18, 24, 32, 33, 39, 41, 61, 69

*Myra Overall 10,317, (14.00), ———, 1, 6, 7, 9, 12, 23, 37, 41, 44, 46, 54, 66

Myth 2837, (14.06), Edward Worth, 15, 65

Myrtle 2d 211, (15.12), Thos. Fitch, 7, 23, 44, 53, 60, 61, 64

Phyllis of Hillcrest 9067, (14.12), W. A. Mullin, 2, 16, 31, 58, 69

Pierrot's Countess 12,480, (14.00), Jos. Kiplinger & Co., 34, 50, 51

Pierrot's Lady Bacon 12,482, (16.10), Jos. Kiplinger & Co., 29, 50, 51, 52, 60

Pierrot's Lady Hayes 11,672, (15.12), Jos. Kiplinger & Co., 7, 34, 50, 51

Pierrot's Picture 12,481, (16.00), Jos. Kiplinger & Co., 7, 34, 50, 51

Pixie 4115, (14.00), Campbell Brown, 12, 14, 37, 49, 50, 51, 52, 59, 72

Plenty 950, (14.08), Thos. T. Turner, 14

Polly Clover 7052, (16.15), Warren F. Daniel, 2, 10, 13, 29, 36, 44, 50, 61, 68

Polly of Deerfoot 15,328, (15.00), E. Burnett, 2, 15, 16, 30

Polynia 10,753, (16.07), I. D. Risher, 6, 14, 21, 24

Pretty 2526, (14.00), J. B. Williams, 2, 29, 57, 64

Pride of Bovina 8050, (16.09), W. S. Rutherford, 5, 21, 24, 29, 30, 44, 70

Pride of Corisande 5323, (16.00), J. H. Walker, 18, 25, 32, 61, 63, 64

Pride of Eastwood 10,227, (20.13), 34

Pride of the Hill 4877, (14.08), G. J. Shaw, 11, 17, 18, 19, 44, 46, 55 .

Pride of Winslow 2613, (14.03), G. Dawson Coleman, (——)

Prince's Bloom 9729, (14.03), E. J. Robbins, 2, 30, 42, 53, 58, 68

Princess 836, (14.12), ——, (——)

Princess 1154, (16.08), ——, (——)

Princess 2d 8046, (46.12½), Com. A. J. C. C., 8, 33, 35, 44, 46. 66, 71, 74

Princess Billworth 6801, (15.10½), Jno. E. Phillips, 2, 13, 20, 30, 34, 42, 56, 58, 67

Princess Bowen 9699, (14.12), Jas. Cloud & Son, 19, 57

Princess Mostar 9700, (17.03), Jas. Cloud & Sons, 12, 19, 25, 33, 51, 52, 57

Princess of Ashantee 13,467, (16.05), S. M. Burnham, 33, 35, 44, 46, 71

Princess of Mansfield 8070, (15.02), A. E. Higley, 29, 50, 51, 52, 60, 63

Princess Rose 6249, (14.08), J. H. Walker, 2, 13, 16, 30, 34, 56, 58

Princess Shiela 7297, (16.04½), G. R. Dykeman, 6, 9, 10, 12, 21, 27, 35, 36, 40

Prize Clementine 10,322, (15.12), J. H. Walker, 41, 46, 65, 67

Prize Rose 13,309, (15.01), W. Simpson, 3, 35

Pulsatilla 2551, (16.03), ——, 4, 14, 30, 37, 38, 49, 58, 64, 70

*Punchinello 11,875, (17.11½), S. M. Burnham, 8, 41, 44, 65, 67, 70

Purest 13,730, (15.04), W. Simpson, 32, 33, 41, 61

Pussie 3035, (19.01), J. H. Walker, 72

Pyrola 4566, (18.06), O. S. Hubbell, 20, 42, 49, 64

Pyrrha 6100, (16.14½), W. Simpson, 4, 6, 33, 57. 74

Queen of Ashantee 14,554, (15.02), S. M. Burnham, 8, 25, 34, 45, 46, 48, 54

Queen of Ashantee 2d 16,657, (14.03½), C. R. C. Dye, 4, 25, 34, 46

Queen of Chenango 17,771, (14.06), J. V. N. Willis, 21, 48, 65, 67

Queen of Delaware 17,029, (18 13), A. Baker, 46, 49, 54, 66

Queen of De Soto 12,318, (14.13), Edward Mayes, 12, 18, 21, 26, 30, 44, 64, 72, 73

Queen Fannie 10,275, (14.02), Hoover & Co., 34, 50, 51

*Queen of Mapledale F. 7036, (14.14), ——, 11, 22, 39, 57, 72

Queen of Prospect 11,997, (14.02), R. S. Kingman, 28, 42, 51, 62

Queen of the North 17,973, (14.00), J. H. Walker, 49

Queensborough 24,345, (17.05), ——, (——)

*Ramilly 17,075, (14.00), ——, 3, 12, 17, 52, 55

Rarity 2d 7724, (14.02), Louis Stracke, 2, 22, 29, 30, 50

Reality 16,537, (15.03½), W. Simpson, 8, 11, 32, 41, 65, 69

Reception 8557, (19.08), W. R. McCready, 29, 46, 47, 65, 66

Reception 3d 11.025, (15.08½), J. H. Walker, 2, 14, 29, 39, 62, 72

Reckless 3569. (17.08), W. Simpson, 12, 14, 53, 54, 65

Regina 32 J. H. B., (20.04), I. of Jersey test, 47.

Regina 2d 2475, (14.08), H. Borden-Bowen, 45, 47, 54

Regina 4th 12,732, (17.13½), H. M. Howe, 47, 49, 54, 55, 66

Renalba 4117, (17.04½), Campbell Brown, 5, 18, 19, 45, 50. 51, 52, 57

Rene Ogden 1568, (15.00), W. S. Taylor, 4, 6, 18, 19, 23, 30, 45, 60, 63

Renini 9181, (14.10½), Chas. Keep, 21, 48, 65

Renown 13,729, (14.06), W. Simpson, 24, 32, 41, 61, 65, 69

Richness 16,536, (14.06), W. Simpson, 32, 33, 41, 61

Rioter Pink of Berlin 23.665, (19.14), Com. A. J. C. C., 8, 18, 37, 66, 71, 73

Rioter's Beauty 14,894, (14.00), V. E. Fuller, 8, 18. 37, 48, 55, 66, 71, 73

Rioter's Nora 21,778, (15.09), V. E. Fuller, 18, 37, 62, 66, 71, 73

Rioter's Ruth 14,882, (14.12), V. E. Fuller, 8, 18, 37, 55. 66, 73

Rioter 2d's Venus 3658, (14.00), ——, 32, 39, 57, 61, 67

Rissa (2173 J. H. B.) 16,014, (19.00), Nathan Brownell, (——)

Robema 3840, (31 days), (54.00), ——, 74

Robinette 7114, (14.01), A. B. Darling, 18, 32, 33, 41, 60, 61, 63

Roland's Bonnie 2d 18,054, (19.02), J. H. Walker, 1, 2, 9, 15, 17, 31, 38, 58, 64

Roll of Honor 13,610, (14.12), J. B. Wilder, 2, 16, 31, 43, 58, 61, 64, 69, 73

Romping Lass 11,021, (15.00), J. B. Wade, 8, 12, 18, 19, 33, 35, 39, 45, 47, 69, 71, 72

Romp Ogden 2d 4764, (15.05), W. E. Oates, 10, 21, 40, 47, 50, 51, 52, 53

Romp Ogden 3d 5458, (14.01), Campbell Brown, 6, 12, 14, 18, 19, 23, 39, 40, 47, 69, 72

Roonan 5133, (18.12), M. C. Campbell, 14, 49, 50, 51, 52, 55, 59

Rosa 663 J. H. B., (18.06), ——, 20

Rosabel Hudson 5704, (15.12), J. H. Walker, 1, 23, 34, 44, 47, 69

Rosa Miller 4333, (17.07), J. H. Walker, 2, 29, 45, 53, 55, 64

Rosa of Bellevue 6954, (18.07½), Thos. H. Malone, 12, 47, 48, 56

Rosa of Glenmore 3179, (17.10), Campbell Brown, 9, 14, 23, 27, 45, 53, 59

Trudie 2d 4084, (15.00), Thos. H. Faile, 6, 24, 32, 61, 67

Trust 23,642, (16.14), ——, (——)

Turquoise 1129, (14.03), John D. Wing, 14, 34

Typha 5870, (16.11), W. Simpson, 3, 18, 32, 35, 57, 61

Uinta 5743, (14.10), M. L. Fink, 4, 16, 37, 70

Undine of South East 4548, (14.00), ——, (——)

Urbana 5597, (16.00), J. C. Johnson, 3, 11, 14, 19, 27, 49, 51, 53, 59

Usilda 832 (14.00), W. S. & H. E. Savage, (——)

Usilda 2d 6157, (15.02½), W. S. & H. E. Savage, 2, 13, 30, 56, 58

Valerie 6044, (15.13), W. B. Montgomery, 2, 9, 11, 29, 30, 43, 44, 49, 55, 68

Valhalla 5300, (17.00), C. P. Markle & Son. 2, 24, 30, 39, 49, 62, 72, 74

Valma Hoffman 4500, (21.09), Sam'l T. Earle, 6, 14, 16, 33, 48, 59

Value 2d 6844, (25.02½), Com. A. J. C. C., 2, 6, 10, 13, 28, 40, 49, 64, 67, 68

Vaniah 6597, (15.09½), T. F. Shotwell, 28

Variella 6337, (18.03¾), W. E. Oates, 4, 11, 14, 36, 57, 59, 70

Variella of Lynwood 10,954, (14.01), M. M. Gardner, 8, 12, 14, 39, 42, 49, 59, 69, 72

Velveteen 7703, (14.13½), Thos. H. Malone, (——)

Verbena of Fernwood 9088 ,(15.00), Harrison Leib, 3, 20

Vesper 1395, (14.00), G. Dawson Coleman, (——)

Vestina 2458, (14.02), W. Simpson, 32, 41, 61

Vespucia 17,455, (14.04), A. J. Fish, 26, 28, 30, 70

Victoria 3175, (16.01), W. L. & W. Rutherford, 44

Victorine La Chaise 2740, (16.00), ——, (——)

Victory 16,379, (15.04½), V. E. Fuller, (——)

Vieva 3d 7642, (16.05), John E. Phillips, 26, 40, 43, 64, 67, 70

Village Maid 7069, (14.00), J. H. Walker, 25, 31, 38, 43, 57

Violet 23, (17.08), ——, (——)

Violet 272 (17.08), J. F. Connelly, 62

Violet 3d 3240, (15.08), T. S. Cooper, (——)

Violet of Glencairn 10,221, (14.04), V. E. Fuller, 9, 22, 28, 36, 42

*Violet of St. Ouens 8626, (17.08), James H. Cushing, 25

Viva Le Brocq 13,702, (17.07), G. B. Smith, 3, 12, 35, 37, 47, 73

Vivalia 12,760, (14.00), Tenn. Breeders' Sale Cattle, 13, 15, 18, 44, 46, 55, 60

Vixen 7591, (17.06), M. M. Gardner, 14, 30, 38, 39, 49, 58, 64, 72

Volie 19,465, (18.01), Henry Sanford, 5, 48, 64, 65.

*Wabash Girl 14,550, (16.00), ——, 2, 20, 22, 31, 38, 49, 54, 62, 68, 69, 72

Wakena 19,721, (16.00), A. P. Foster, 15, 22, 24, 28, 52, 68

Walkyrie 5708, (14.01), H. M. Howe, 5, 24, 31, 34, 42, 46, 47, 54, 55, 66

Warren's Duchess 4622, (16.01), C. Bordwell, 26, 53, 62

Webster's Pet 4103, (14.02), W. H. Walrath, 6, 10, 14, 21, 35, 36, 40, 50, 59, 71, 74

Welma 5942, (17.08), Ariel Low, Jr., 11, 20, 39, 54, 72

White Clover Leaf 4512, (17.15), J. H. Walker, 13, 30, 42, 53, 58, 64

White Frost 17,431, (16.02), ——, (——)

Willis 2d 4461, (16.03), G. Dawson Coleman, 12, 28, 54, 65

Witch Hazel 1360, (14.00), Thos J. Hand, 11, 12, 72

Witch Hazel 4th 6131, (15.05½), Campbell Brown, 35, 36, 54, 63, 72

Woodland Lass 3444, (14.00), J. H. Walker, 1, 23, 31, 38, 43

*Woodland Margaret 6215, (14.10½), ——, 11, 12, 30, 39, 51, 52, 72

Wybie 595, (17.04), O. S. Hubbell, 10

Yellow Locust 10,679, (14.10½), H. M. Howe, 8, 24, 28, 31, 32, 38, 50, 58, 70

Young Duchess 497, (15.08), J. H. Walker,

Young Fancy 9032, (17.00), Isaac W. Stokes,

Young Garenne 13,641, (17.08), ——, 8, 12, 34, 44, 46, 47

Zalma 8778, (15.05), W. Simpson, 4, 18, 31, 32, 41, 61

Zampa 2194, (18.00), ——, 6, 14, 16, 32, 51, 61

Zina 1434, (14.00), ——, 6, 14, 21, 64

Zittella 2d 11,922, (17.08½), W. Simpson, 8, 11, 32, 33, 41, 61

Zithey 9184, (16.07), O. S. Hubbell, 21, 42, 48, 49, 65

,CRESCENT FARM HERD.

JERSEY CATTLE.

SERVICE BULLS:

TRAILER 7160.

Full brother in blood of Bomba 10,330 (21 lbs. 11¼ oz., official record), their sires being the same, and their dams full sisters. Average 7 days' record of dam and sire's dam, 18 lbs. 13 oz. Service fee, $50. Sons of Trailer taken at 4 weeks old as full pay for a second service.

CRESCENT TRAILER 12,771.

Son of Trailer 7160, and Doe Magna 7122, a sister of Gilderoy, who tested 13 lbs. 8 oz. with only 2 sound teats. Service fee, $50. A heifer calf guaranteed or the service fee refunded, and 2 free services given besides.

MARIUS ALBERT 14,542,

40⅘ per cent. Albert 44, backed by Lady Mary 1148 and Pansy 6th 38, the combination which produced Signal. Service fee,. $25.

Young stock for sale. The herd is made up of Rioter-Alphea, Albert, St. Helier, Rajah and Gilderoy blood. Address

F. M. CARRYL,

Passaic Bridge, N. J.

Only 10 Miles from New York on Erie Railroad.

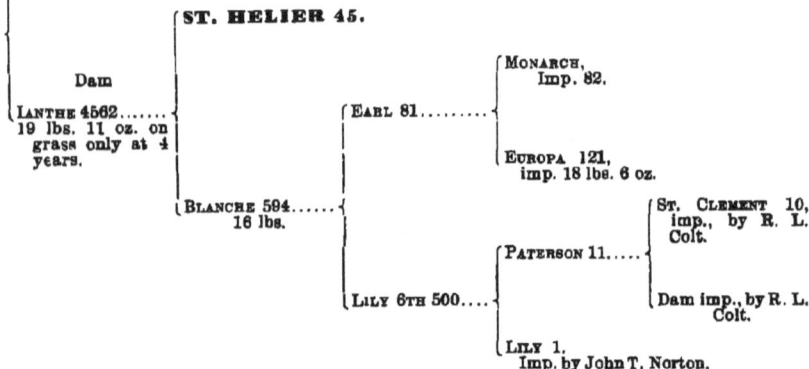

OAKRIDGE HERD.

JERSEY CATTLE

J. B. Aycrigg, M. D.,

PASSAIC BRIDGE, N. J.

THE SIMPSON HERD

OF

JERSEY CATTLE

IS THE LARGEST HERD IN THE WORLD.

The following Bulls are in use; for records of their families, see preceding pages:

MERCURY 432.	EDDINGTON 2250.
YOUNG MERCURY 7485.	REX 1330.
MERCURY, JR., 7490.	FARMER'S GLORY 5196.

&c., &c.

Choice young Bulls and Heifers of the best butter blood for sale at reasonable prices. Address

WILLIAM SIMPSON,

51 Chatham Street, New York.

www.ingramcontent.com/pod-product-compliance
Lightning Source LLC
Chambersburg PA
CBHW021951190326
41519CB00009B/1211